Discrete Mathematics With Logic

Discrete Mathematics
With Logic

Martin Milanič

Brigitte Servatius

Herman Servatius

ACADEMIC PRESS

An imprint of Elsevier

ELSEVIER

Academic Press is an imprint of Elsevier
125 London Wall, London EC2Y 5AS, United Kingdom
525 B Street, Suite 1650, San Diego, CA 92101, United States
50 Hampshire Street, 5th Floor, Cambridge, MA 02139, United States
The Boulevard, Langford Lane, Kidlington, Oxford OX5 1GB, United Kingdom

ISBN: 978-0-443-18782-7

For information on all Academic Press publications
visit our website at https://www.elsevier.com/books-and-journals

Publisher: Katey Birtcher
Editorial Project Manager: Ali Afzal-Khan
Publishing Services Manager: Shereen Jameel
Production Project Manager: Gayathri S.
Cover Designer: Matthew Limbert

Typeset by VTeX

Working together
to grow libraries in
developing countries

www.elsevier.com • www.bookaid.org

Dedication

is here encoded, but not encrypted:

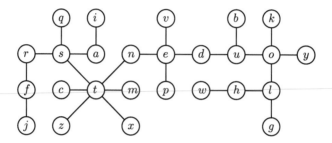

Contents

4.   Formal logic

5.   Induction

6.   Set structures

Preface

This book is designed to be read by modern students.

This book is neither an encyclopedia of discrete mathematics, nor a dictionary. The text is written to be read in support of the accompanying exercises, and the exercises have been written specifically to illustrate key definitions and techniques of discrete mathematics, particularly those of interest to computer scientist majors.

The topics have been chosen to make up a course in Discrete Mathematics for first or second year students of mathematics or computer science. The essential chapters are Chapter 2, on basic set theory, Chapter 4, on basic formal logic, and Chapter 5 on induction. These chapters contain the three proof forms which students must write examples of in our course, "double inclusion", "double implication", and induction. They produce these either individually or, for a large class, in groups. The essential material in Chapter 1 is at the high school level, but experience says it is best not to skip it.

Every chapter has at least six main sections, the optional ones marked with a †. Each main section contains three exercises specifically designed to reinforce the definitions in the section. There are always exactly three because even a moderately curious student seeing three exercises will try to do them. Most students encountering a long list of exercises will not try a single one, and will instead wait for an assignment. Interspersed in the text one finds the symbol ✠, which is a signal that the reader should stop reading and think before continuing.

The main sections have all the material that the student needs to understand for the quizzes and tests.

Following the main sections are sections labeled "Case studies". Case studies use and expand upon the material in the main sections, or anticipate ideas which come later. Some introduce new material, some focus on interesting problems, some provide proofs of the more difficult theorems, but only if the proof illustrates important ideas. We avoid proofs for proof's sake. Case studies have no exercises. All case studies are optional both from the point of view of the teacher, and the instructor. While it is possible to skip all the case studies, we believe that it is an important aspect of the course to have discussions about topics without the distraction of homework, grades, and tests. The summary exercises at the end of the chapter never refer to the case studies, and the list of "concepts that should have been learned" never refer to them either.

The case studies have been placed so that, by the time they appear, all background concepts have been covered, but many work well in earlier or later chapters, and covering them in between main sections is just as effective as waiting until the main material is done. Some are very easy, some are more challenging, and all are to some degree open ended. The case studies also work very well for individual or group student presentations.

The book contains few citations and almost no bibliography. Students should have no trouble finding sources for the material presented if desired. On the other hand, for many problems and results we have deliberately altered the names and the framing (e.g., "Euclid's Coin Problem") so that the student cannot easily just "google it", and will perhaps think it through on their own instead.

The pace of the class should be to cover two sections, main or case studies, in each 50-minute class period.

For our students we always use all five of the initial chapters, varying only the choice of case studies. On average we use two case studies per chapter, either those in the text or others written in the same format, the choice depending on class interest. The last week or two is devoted to some selection of the later material.

Chapter 1

Discreteness

1.1 What is discrete mathematics?

Defining discrete mathematics is not as easy as defining many other branches of study. In biology you study living things, in astronomy you study heavenly bodies, in vector calculus you study calculus on vectors. For discrete mathematics, the question is not only *what* is being studied, but *how* it is being studied. The easiest way to answer the question "What is discrete mathematics?" is to consider a variety of problems, mathematical and non-mathematical, and to examine for each of them to what extent we would consider it to be part of discrete mathematics.

Problem list

1. Maximize $x^3 - 100x$ on the interval $[-5, 21]$.
2. Is $\sqrt{2}$ rational?
3. *Pythagoras's problem*: Show that for any right triangle, the square of the hypotenuse is equal to the sum of the squares of the other two sides.
4. *The Nantucket Map Problem*: There is a large map of Nantucket pinned to a bulletin board. Someone pins a small map of Nantucket on top of the large map. Find a point to place a pin so that it marks the same location on both maps.
5. *The Wolf-Goat-Cabbage Problem*: Can a man transport three items, a wolf, a goat, and a basket of cabbages, across a river using a boat which is only large enough for him to transport one item of cargo at a time, and keeping in mind that, if left unguarded, the wolf will kill the goat and the goat will eat the cabbage.
6. *Albertson's Magic Trick*: An MC asks a victim from the audience to pick a number from 1 to 31 and then answer truthfully whether or not it lies on each of the following five cards.

16	18	20	22	8	10	12	14	4	6	12	14
17	19	21	23	9	11	13	15	5	7	13	15
24	26	28	30	24	26	28	30	20	22	28	30
25	27	29	31	25	27	29	31	21	23	29	31

Discrete Mathematics With Logic. https://doi.org/10.1016/B978-0-44-318782-7.00006-X

2	6	10	14	1	5	9	13
3	7	11	15	3	7	11	15
18	22	26	30	17	21	25	39
19	23	27	31	19	23	27	31

Given the answers, how can the MC determine the number?

7. *The Presidential Problem*: What was George Washington really like?

First of all, which of these problems is even mathematical? Clearly the first three problems are. The others seem questionable. Albertson's Magic Trick has numbers in it, but are they actually part of the problem? Would the trick be any different if, instead of numbers, the cards contained 31 cartoon characters? On the other hand, the Nantucket Map Problem has no numbers at all, but the solution might involve geometry, which is certainly part of mathematics. The rest of the problems on the list seem to be essentially non-mathematical.

What about discreteness? Most people use the word discreet in the sense of *tactful*, or *unobtrusive*.[1] That is not our meaning at all. In this subject, *discrete* is to be understood in the sense of separated or granulated, as opposed to that which is smooth or continuous. In electronics, a digital signal would be considered discrete, whereas an analog signal would be not. We think of a box of billiard balls as a discrete collection of objects, but a bucket of water as containing a non-discrete fluid. What, you might ask, about a bucket of sand? Is it discrete, continuous or somehow lying on the boundary? On the one hand, we can draw on the surface with our fingers, on the other we can pick out individual grains with tweezers. If we are pouring sand, it seems continuous, acting as a fluid, but once poured, it doesn't flow into a puddle, like water, but forms a heap whose shape is determined by the size and type of grains. So which is it?

In this class, we will have many mathematical definitions. Learning how to manage mathematical definitions is a key skill to be developed, but discreteness has not been defined mathematically. Discreteness is an important idea, one that shapes our subject, but deciding whether or not a particular idea or phenomenon is discrete might depend on your experience or point of view. It is for this reason that this first section of the text seems more philosophical, than mathematical.

Let's try to apply this idea to the numbers we know. On the one hand, we have the *natural numbers*, $\mathbb{N} = \{0, 1, 2, 3, 4, \ldots\}$, which are perfectly appropriate for dealing with discrete objects, such as counting the number of billiard balls, and the related *integers*, $\mathbb{Z} = \{\ldots, -3, -2, -1, 0, 1, 2, 3, \ldots\}$. More generally we have the real numbers, $\mathbb{R}$, which include $.5$, π, $-e$, and in fact all numbers with a decimal representation, that is, which lie on the continuous number line.

$$\text{HHH HHH //} \quad \text{vs.} \quad \underset{-2 \quad -1 \quad \ 0 \quad \ 1}{\longmapsto\!\!\!\!\!\longrightarrow}$$

---

[1] Discrete mathematics is using your calculator on a test when the proctor is not looking.

So the first problem on our list, optimizing a continuous function of a continuous variable, is definitely a problem in continuous and not discrete mathematics.

The second problem involves rational numbers, $\mathbb{Q}$, numbers whose decimal representations are finite or repeating. At first glance the rational numbers don't seem quite so discrete as the integers since they are packed so close together on the number line, but since every rational number is associated with just two integers, the numerator and the denominator of a fractional representation, it seems we must classify the rational numbers as discrete. Problem 2 requires us to show that $\sqrt{2}$ is irrational. To do this directly looks intimidating, since it seems impossible to first generate the entire decimal representation of $\sqrt{2}$ and then show that it is infinite and non-repeating. If however, we hadn't been given the hint in advance that $\sqrt{2}$ is irrational, we might try instead to show that it *is* rational, for instance by finding p and q so that $\sqrt{2} = p/q$. That, at least, looks like discrete mathematics. To make sure there is exactly one p and q to be found, let's assume that p/q is in lowest terms, so p and q have no common factor. So set $\sqrt{2} = p/q$ and let's try to solve. Squaring gives, $2 = p^2/q^2$, or $2q^2 = p^2$. So p^2 must be an even number, hence p itself must be even. So we can set $p = 2k$ and we have $2q^2 = (2k)^2 = 4k^2$, or $q^2 = 2k^2$, so q^2 is even and hence q itself must be even. But now p and q both have a common factor, 2. This is a contradiction. Hence we conclude $\sqrt{2} = p/q$ has no solutions with integers in lowest terms, so $\sqrt{2}$ is irrational. The second problem is solved, and even more, the problem was solved discretely. We used only simple algebra, properties of that very useful number 2, and logic. So it seems we should classify problem 2 as belonging to discrete mathematics.

But wait, if we think back to the calculus problem that we rejected earlier as non-discrete, its solution actually looks similar. There the problem is defined by just a few numbers: 3, 100, -5 and 21, and the solution involves computing the derivative. The derivative is computed by just pushing the symbols around according to a prescribed method, e.g., x^3 turns into $3x^2$. Although the problem is continuous, the solution is via a sequence of separate distinct operations. Indeed the word *calculus* comes from the Latin word for pebble, with the idea that solving problems in differential calculus is analogous to counting pebbles out of a bowl. So there seems to be something discrete about the first problem after all.

The wolf-goat-cabbage story has both continuous and discrete aspects but, according to the rules of the problem, the continuous aspects (the weight of the boat, the speed of the river, or the height of the man) are not relevant to the solution. Solving the problem quickly reduces to considering a discrete collection of situations, or "states", each indicating on which side of the river the man, goat, wolf, and cabbage are. So at least Problem 5 is a discrete problem, and you should be easily able to solve it by trial and error, perhaps on a table with five coins to represent the characters.

Let's now consider the magic trick. We will violate the fundamental rule of magic and reveal the secret of the trick. Each card has an active region:

16	18	20	22
17	19	21	23
24	26	28	30
25	27	29	31

8	10	12	14
9	11	13	15

4	6
5	7

2
3

1

The numbers have been placed on the cards so that, for the first card for which the answer is yes, the secret number will not only be on that card, but in the active region. So if the answers are no, no, no, no, yes; then the active region on the fifth card is only a single number, and the secret number is 1. Each subsequent answer allows us to divide and shrink the active region in half, until at the end it contains only the required number. Each time the region is shrunk, it is shaped like the next active region. The secret number will be on the left (top) half if the answer is no, and on the right (bottom) half if the answer is yes. Shrinking the active region in this way will always yield a single number by the fifth question.

Let's try it out. If the number is 6, the answers would be: no, no, yes, yes, no. So the active region starts out as $\begin{array}{cc} 4 & 6 \\ 5 & 7 \end{array}$ for the yes to question 3, and then shrinks to the right half $\begin{array}{c} 6 \\ 7 \end{array}$ for the next yes, and to the top $\boxed{6}$ for the last no. You should try the secret method with some other numbers to convince yourself that it works. With just a little practice as MC you can give the answer before the victim has even finished saying the last answer. That's the magic. Nobody would be impressed if the MC wrote down all the answers, thought for a few minutes, and then answered correctly. Now, to decide if the method works, we just have to try it with all thirty-one numbers. If it doesn't work, we are done, but what if it does work? We will still not know *why* it works, and have no clue how the trick was ever designed in the first place.

So, just as with Wolf-Goat-Cabbage, we have a discrete problem, but is it a problem in discrete *mathematics*? So far, the answer for both is *no*. What is missing is the mathematical analysis of the problems. Just playing around with discrete objects does not constitute mathematics, whether the objects are numbers or not. And mathematical analysis is to be meant in the most general sense, as in Problem 2, whose solution involved only evenness, oddness, and logic. To start the analysis, in order to make these problems mathematical, we should ask and answer some mathematical questions about these problems.

Let's start with Albertson's Magic Trick. There are many questions we could ask about whether it works, or why it works. For instance, can we add some new numbers to the cards? Or perhaps we want to make a version of the trick with the 31 numbers replaced with the 50 states. Is that possible? One restriction is that, no matter how we arrange the 50 states on the cards and no matter what clever active regions we have come up with, if the victim has the same five answers for Illinois as for Hawaii, say, (yes, yes, no, yes, no), then it will be impossible for the MC to decide correctly between them. So the number of states we can place on the cards to make an effective trick is restricted by the number of answer sequences. How many are sequences are there? There are two possible answers for the first question, and for *each of these* there are two possible answers for the second question $2 \cdot 2 = 4$, and for each of these there are two answers for the third question $(2 \cdot 2) \cdot 2 = 2^3$ and so on, giving us $2^5 = 32$ answer sequences. So a version with all 50 U.S. states is impossible with just five cards and yes-no questions.

Does this show that the trick works? No. We have not counted the yes-no sequences *in the trick*, just the number of sequences in general – the sequences you would get if you didn't show the victim the cards at all but just asked random questions. So all we have shown is that the trick is *plausible*, that it is not impossible from this point of view. Even though we have not solved the problem, we are now applying mathematics to the discrete problem, and that is discrete mathematics.

We can use this sequence counting idea, called the *Multiplicative Principle*, for the Wolf-Goat-Cabbage problem as well. How many different states are there in the problem? For each object in the story we can ask the question "Which side of the river is it on?", and record the answers to the questions. So for man, the boat and the three articles of cargo we have 2^5 possible states, by the same principle we used before. Does this solve the problem? No. But the analysis has started, and you may be getting ideas on how to attack it.

Finally, consider the Presidential Problem which we found to be neither discrete nor mathematical, and so the least likely of any of the problems to be of interest in a course of Discrete Mathematics. The question concerns George Washington's character. Game developer Gary Gygax described the following system to mathematically classify and analyze character, dividing the concept into six major traits, to be assigned a number from 3 to 18, (since in the game the character traits could be assigned by rolling three dice), and two alignments allowing three states each, see Table 1.1. This system reduces the Presidential Problem to determining to which of the (multiplicative principle yet again!) $16^6 \cdot 3^2 = 452,984,832$ Gygax character types George Washington belongs.

In summary, it seems that any problem can be approached mathematically if we are willing and able to define and analyze it rigorously, and that is what we must do if we are ever going to really solve it. In addition, the fact that we have

TABLE 1.1 Dungeons and Dragons character discretization.

Strength	Intelligence	Wisdom	Constitution	Dexterity	Charisma
3-18	3-18	3-18	3-18	3-18	3-18
Alignments					
		Good-Neutral-Evil	Lawful-Neutral-Chaotic		

a solution means we have reduced the problem somehow to a sequence of steps, calculations, or procedures; that is, something discrete. So it is not surprising that discrete mathematics almost always appears in the end-game of a solvable problem; that discrete mathematics therefore is a gigantic subject; that it is found throughout our technological society. That said, all our problems do not end in discrete mathematics, by any means, since most of the important problems in life are not solvable, and are the province of poets, artists, philosophers, etc.

Exercises

1. Order the problems in the list discussed in this section from least to most discrete.
2. Consider methods people have of communicating. Identify three which are essentially discrete.
   Identify three which are essentially non-discrete.
   Justify your responses.
3. Consider the following outdoor activities. Order them from least to most discrete: jump-rope, frisbee, hopscotch, pogostick. Justify your ordering in a few words.

1.2 The Multiplicative Principle

Sequences of independent choices

In Section 1.1 we used the multiplicative principle three times as a tool to analyze a problem. So we state it formally here:

Theorem 1.1 (The Multiplicative Principle). *The number of sequences of n independent choices, $(c_1, c_2, c_3, \ldots, c_n)$, where the number of options for choice c_i at stage i is O_i is the product of all the numbers of options:*

$$O_1 \cdot O_2 \cdot \cdots \cdot O_n$$

if the choices are either independent or weakly independent.

The result seems clear, but when we applied it earlier no mention was made of independence.[2] If you have taken a course in probability, then the independence idea is essentially the same. We say that the kth choice is *independent* of the earlier choices if there is no sequence of choices $c_1, c_2, \ldots, c_{k-1}$, such that the kth choice is at all restricted from the full O_k options.

Were the choices in the Magic Trick independent? You would have to look carefully at the cards to tell. But before you peak, just suppose that every number occurring on the fourth card occurred at least once on one of the three previous cards. Then the fourth choice would not be independent because it would not be possible, after answering (no, no, no, _, _) for the first three cards, to answer yes for the fourth card. In that situation, the choices would not be independent.

For the actual cards, we know that the answers must be dependent since the victim chooses a number from 1 to 31, but the multiplicative principle predicts $2^5 = 32$, so some sequence of choices does not occur.

In the absence of independence (or weak independence), the multiplicative principle merely gives us an upper bound on the sequence of choices.

Exercises

1. An online order form for a winter coat allows you to choose up to four colors for the coat, and four colors for an optional hood, and either ordinary or Red-Sox themed buttons.
   First give a guess as to the number of types of coats which can be ordered.
   Then compute the number of different types of coats which can be ordered

2. You are in a restaurant and order the special, which has three choices of appetizer, one of which is a salad with three choices of salad dressing. There is a vegetarian option to the entrée and three choices of dessert.
   Would you guess that ten different meals are possible?
   Compute how many different meals can be ordered. Justify your response.

3. It is your first day of class and you have to fill out four feedback surveys originating from four different administrative offices, to measure your happiness and efficiency.
   The first survey has six true/false questions. The second and third surveys were obviously edited with the same survey composition app, and both have large animated smiley faces which read aloud to you twelve multiple choice questions whose four alternatives are labeled comfortingly "chocolate", "sunsets", "cuddle", and "joy". The fourth survey is exactly the same as the first survey with the order of the questions shuffled – but it must be filled out anyway.
   How many different survey responses are possible?
   The student must complete all surveys and nothing can be left blank.

---

[2] We will discuss weak independence in Section 1.3.

1.3 Binomial coefficients

Weakly independent choices

Suppose you have 15 billiard balls and want to place them in the rack. How many ways are there to do that? The balls are numbered from 1 to 15, and the rack is just an oriented triangular array.

$$
\begin{array}{ccccccc}
 & & & 5 & & & \\
 & & 6 & & 9 & & \\
 & 10 & & 11 & & 1 & \\
3 & & 15 & & 2 & & 14 \\
7 & 4 & & 8 & & 12 & 13
\end{array}
$$

Imagine placing the balls one by one from top to bottom, since any ball can be in any position, it is tempting to take the number of options for the ith ball to be $O_i = 15$. We will then get 15^{15} which will be a very generous upper bound since the choices are not independent. Only the first placement has the full 15 options since no ball can be placed twice.

But it is not hard to see what to do next. Reset the number of options to take into account the decreasing range of choice

$$O_1 = 15, \quad O_2 = 14, \quad O_3 = 13, \quad \ldots, \quad O_{14} = 2, \quad O_{15} = 1.$$

We have almost solved the problem, but there is an interesting subtlety. Once, say, ball 5 is chosen for the first position, that ball is no longer an allowable choice to be placed in any of the other positions. But it *would* have been available if a different ball had been chosen initially. The available second choice is dependent on the first one, and so are all the subsequent ones. On the other hand, although the particular choices change, the *number*, O_i, of subsequent options at each stage does not. Regardless of which ball was chosen first, there is the same steadily decreasing number of choices after the first one; so by the same argument we used before, the number of ways to rack the balls is the same as the number of sequences of allowable choices, $15 \cdot 14 \cdot 13 \cdots 2 \cdot 1 = 15!$.

We say that the kth choice is *weakly independent* of the earlier choices if, regardless of what has been chosen so far, there are always O_k options available. So the calculation above actually does follow from the Multiplicative Principle (Theorem 1.1) as stated in the previous section.

Theorem 1.2. *The number of ways to order n objects, with $n > 0$, is $n!$.*

The factorial is defined by the familiar equations

$$
k! = \begin{cases} 1 & k = 0 \\ (k)(k-1)\cdots(1) & k > 0 \end{cases} \tag{1.1}
$$

for $k \geq 0$. The fact that $0! = 1$ looks peculiar, but is absolutely necessary.

Choosing

Suppose we have a class of 11 kindergartners, and want to choose five of them to take a bunch of flowers to the principal. How many ways are there to do that? The teacher asks for volunteers, and all the kids raise their hands and wait expectantly to be chosen. The teacher chooses the first kid, 11 choices, then for each of these a second, weak independence giving $11 \cdot 10$, then a third $11 \cdot 10 \cdot 9$, a fourth $11 \cdot 10 \cdot 9 \cdot 8$, and finally a fifth $11 \cdot 10 \cdot 9 \cdot 8 \cdot 7$. In particular, for little Julie, it is very important to her not only to be chosen, but to be chosen *first*. If she is not among the first three chosen, she might even throw a tantrum, and not even be comforted if she is later selected fourth or fifth.

When the kids take the flowers down to the office, the principal notices of course *who* has been chosen to bring the flowers, but has no interest in the order in which they were chosen. All the 5! ways which the teacher had of choosing those five kids, so important to little Julie, would be considered the same to the principal.

So we have two answers to the counting problem. From Julie's point of view, there are

$$11 \cdot 10 \cdot 9 \cdot 8 \cdot 7 = \frac{11!}{6!}$$

ways of choosing, in which the actual process of choosing is important, and for the principal there are

$$\frac{11 \cdot 10 \cdot 9 \cdot 8 \cdot 7}{5!} = \frac{11!}{5!6!}$$

ways in which the manner of choosing not considered, only *who* was chosen. The first is often called the *ordered selection*, and the second the *unordered selection*. There is a special notation for the unordered selection, in this case it would be $\binom{11}{5} = 11!/5!6!$. Other notations, such as C_5^{11}, are sometimes seen.

The numbers $\binom{n}{k}$ have been around for a very long time, and have been re-discovered in many different contexts, so the notation and language which has evolved around them is a hodgepodge of that history. The number $\binom{n}{k}$ is most often called a *binomial coefficient*, and the particular one referred to here is read off as "n choose k". That is often the source of many student, and non-student, errors since the mathematical language does not mesh with the natural language we use to express the problems. So Julie, who cares so much about the choice, does not use $\binom{11}{5}$ to count from her point of view, but the other one. In English you can get over this difficulty by, just to yourself, thinking of $\binom{11}{5}$ not as "n choose k", but "n grab k", since 'grabbing' has much more the connotation of selecting altogether than 'choosing'. (But don't expect that if you say "n grab k" that anyone else will know what you are talking about.)

Here is a summary:

Theorem 1.3 (Counting Selections). *The number of ways to perform an unordered selection of k objects from n objects is $\binom{n}{k}$.*
The number of ways to perform an ordered selection of k objects from n objects is $\binom{n}{k}k!$.

Example 1.4. How many four-letter words have two vowels?

First, by vowel is meant one of the five letters a, e, i, o, and u. Also, for problems of this type, we are not to be hunting through a dictionary to determine actual words, so *lite* and *xoox* count. We have four letter positions _ _ _ _ and we want to choose letters to fill them. What you don't want to do is to let the word "choose" prompt you to take a binomial coefficient prematurely, without thinking first if that is the appropriate measurement. Also, it is often a mistake to start "choosing" before considering how those choices will combine with other parts of the problem. Here, the problem is very difficult to approach strictly left to right, since the choices in method are dependent. Actually, it is best not to start choosing any letters, but rather the positions to be occupied by vowels. For this one chooses (grabs) two of the four positions in $\binom{4}{2}$ ways, say the 1st and 4th, ___.

The word to be constructed has now been organized for the placement of letters, but is still empty. Note that, as required by the multiplicative principle, the number of choices we make in now filling in the word does not depend on which organization we have specified. Now select vowels for the two vowel positions, left to right, say e a, and consonants for the two consonant positions, say $etba$, and we are done. We have determined $\binom{4}{2}5^2(26-5)^2$ possibilities.
◇

Exercises

1. How many 10-digit decimal numbers (zeros in front ok) have at least five 1's?
2. How many 10-digit decimal numbers (zeros in front ok) have exactly five even digits (0, 2, 4, 6, 8)?
3. There is a drawer with three compartments, one filled with oil, one with water, and one with gold paint. 15 billiard balls are to be distributed into the drawers. How many ways are there to do this?

1.4 Pascal's Triangle

The binomial coefficients may be arranged in a convenient pattern:

$$\binom{0}{0}$$

$$\binom{1}{0} \qquad \binom{1}{1}$$

$$\binom{2}{0} \qquad \binom{2}{1} \qquad \binom{2}{2}$$

$$\binom{3}{0} \qquad \binom{3}{1} \qquad \binom{3}{2} \qquad \binom{3}{3}$$

$$\binom{4}{0} \qquad \binom{4}{1} \qquad \binom{4}{2} \qquad \binom{4}{3} \qquad \binom{4}{4}$$

$$\binom{5}{0} \qquad \binom{5}{1} \qquad \binom{5}{2} \qquad \binom{5}{3} \qquad \binom{5}{4} \qquad \binom{5}{5}$$

$$\binom{6}{0} \qquad \binom{6}{1} \qquad \binom{6}{2} \qquad \binom{6}{3} \qquad \binom{6}{4} \qquad \binom{6}{5} \qquad \binom{6}{5}$$

$$\vdots$$

which is called *Pascal's Triangle*. It is important to note that we will follow common computer science practice and label ordered objects starting from 0. So the entry $\binom{4}{2}$, which you may regard as occupying the 3rd position in the 5th row, we prefer to regard as being in the 2nd position of the 4th row, which fits the notation much better. We do have to get used to the fact that the top entry is the whole 0th row, and that the nth row has $n + 1$ entries, from the 0th, $\binom{n}{0}$, to the nth, $\binom{n}{n}$.

Just as important as the rows, the entries of Pascal's Triangle are organized in forward diagonals, the "choose 0's", the "choose 1's", the "choose 2's", etc. The 0th diagonal is entirely made up of 1's, since $\binom{n}{0} = 1$, and the first diagonal identifies the row number, since $\binom{n}{1} = n$ – except for the 0th row, which has no first entry.

If we look at Pascal's Triangle with the binomial coefficients computed, we have this familiar array of numbers

$$
\begin{array}{ccccccccccccc}
 & & & & & & 1 & & & & & & \\
 & & & & & 1 & & 1 & & & & & \\
 & & & & 1 & & 2 & & 1 & & & & \\
 & & & 1 & & 3 & & 3 & & 1 & & & \\
 & & 1 & & 4 & & 6 & & 4 & & 1 & & \\
 & 1 & & 5 & & 10 & & 10 & & 5 & & 1 & \\
1 & & 6 & & 15 & & 20 & & 15 & & 6 & & 1
\end{array}
$$

in which you must not fail to identify the 20 at the bottom as $\binom{6}{3}$ not $\binom{6}{4}$, since it is only the 3rd entry on that row, "counting from 0". The computed triangle also makes plain the obvious mirror symmetry, because the number of ways to grab k things from n things is exactly the same as the number of ways of letting $n - k$ things remain ungrabbed:

$$\binom{n}{k} = \binom{n}{n-k}.$$

It is also easy to show that the entries in each row increase to the middle, and then decrease:

$$\binom{n}{k} < \binom{n}{k+1} \quad \text{for } k \leq (n-1)/2 \tag{1.2}$$

Computing the entries may seem tedious, since it involves so many multiplications, and the numbers get big rather quickly. (How quickly? — Later.) But there is an important shortcut for generating them. Each entry in Pascal's triangle which is not on the boundary, is the sum of the two entries right above it, one to the left, and one to the right. For a binomial coefficient not on the boundary, both entries are at least 1, so let's write it as $\binom{n+1}{k+1}$ The entry above $\binom{n+1}{k+1}$ and to the right is on the same forward diagonal, and so is in the same *choose group*, so is $\binom{n}{k+1}$, and the one to the left must be $\binom{n}{k}$; so we are claiming that

$$\binom{n}{k} + \binom{n}{k+1} = \binom{n+1}{k+1}. \tag{1.3}$$

There are several proofs of this key result. The one which follows now was my professor's favorite: For $\binom{n+1}{k+1}$ you want to grab $k+1$ objects from $n+1$ objects. Assume that one is precious, that it is made of gold. Either you grab the golden object, or you don't. There are $\binom{n}{k+1}$ ways to avoid the golden one. On the other hand, once you have the golden one, there are $\binom{n}{k}$ to grab the ordinary ones.

The preceding four sections are taken from a branch of discrete mathematics called **combinatorics**. It is very useful, and your university probably has several courses in just that.

Exercises

1. Write out the first 8 rows of Pascal's triangle.
   Circle $\binom{7}{2}$, $\binom{7}{5}$, and $\binom{8}{4}$.
2. Show that $\binom{100}{49} < \binom{100}{50}$.
   [Hint: decide if the ratio is larger or smaller than 1.]
   What about $\binom{100}{50} < \binom{100}{51}$?
3. Check that

$$\binom{8}{4} = \binom{4}{4}\binom{4}{0} + \binom{4}{3}\binom{4}{1} + \binom{4}{2}\binom{4}{2} + \binom{4}{1}\binom{4}{3} + \binom{4}{0}\binom{4}{4}$$

Can you find an explanation for this "coincidence"?
Can you find a similar statement for $\binom{12}{6}$?

1.5 Binary numbers

In discrete mathematics, and in computer science, it is often convenient to use the numbers expressed in other bases than our usual one, base 10. Just to recall, a decimal number like 1776 is represented by just four digits, the "digit" coming from the Latin word for finger, by the clever system of scaling each of the digits by a power of 10:

$$1776 = 1 \cdot 10^3 + 7 \cdot 10^2 + 7 \cdot 10^1 + 6 \cdot 10^0.$$

The number 10 is not a digit in the decimal system, but 0 is.

Analogously, in the binary system, the number 2 is not a binary digit, or *bit*. The only bits are 0 and 1. A binary number is represented by a sequence of zeros and ones, like

$$110,010,111 = 1 \cdot 2^8 + 1 \cdot 2^7 + 0 \cdot 2^6 + 0 \cdot 2^5 + 1 \cdot 2^4 + 0 \cdot 2^3 + 1 \cdot 2^2 + 1 \cdot 2^1 + 1 \cdot 2^0.$$

There is no standard about placing commas, and they are usually left out. Sometimes one adds a subscript, in this case 2, to indicate the number base of the representation, especially if the base is not clear from the context: $8_{10} = 10_8 = 22_3 = 1000_2$, which is somewhat inconsistent since the subscript is usually still in base 10, but no system is perfect.

If you have never done so, it is instructive to list the 16 four-digit binary numbers from 0000 to 1111 in order, and observe the developing pattern of 0's and 1's, comparing it to what you would expect from the list of base 10 numbers.

You have probably heard that computers are fundamentally binary because 0 and 1 can represent "off" and "on". And that still by and large is true, since electrical circuits can easily be built like that, but many other architectures are also possible. But even without that consideration, binary numbers would still be of interest because of other fundamental dichotomies like "no" and "yes", "negative" and "positive", "false" and "true", etc.

For example, binary numbers play a fundamental role in Albertson's Magic trick, one of the problems we considered in Section 1.1. The "magic" was in associating the sequences of yes's and no's for the five cards with the sequences of bits of the binary numbers from 00001_2 to 11111_2. So the number $25_{10} = 11001_2$ would be associated with the object pictured on the zero'th, third, and fourth card, but not on the first and second. The trick as presented takes as objects the numbers themselves, undisguised beyond that they are represented in base 10. The zero'th card consists of all five bit numbers whose zero'th bit is 1, the first card those whose first bit is 1, etc.

One can also do arithmetic on binary numbers. Binary numbers are added and multiplied in the same way as decimal numbers, yet with many fewer digit facts to be known, but many more carries. So $13 + 7 = 20$ and $5 \cdot 5 = 25$ would be done as follows:

$$\begin{array}{cccc}
1^1 & 1^1 & 0^1 & 1 \\
+ & 1 & 1 & 1 \\
\hline
1 \quad 0 & 1 & 0 & 0
\end{array}$$

$$\begin{array}{ccccc}
 & & 1 & 0 & 1 \\
\times & & 1 & 0 & 1 \\
\hline
 & & 1 & 0 & 1 \\
 & & & & 0 \\
1 & 0 & 1 & & \\
\hline
1 & 1 & 0 & 0 & 1
\end{array}$$

While it is true that the number bases 2 and 10 are by far the most popular, a facility with others is often quite handy, or even essential.

Exercises

1. Let $n = 1101101101$ and $m = 1110110110$, be binary numbers. Compute $n + m$, $m - n$, $n/2$, and $n^2 + 1$ in binary format. Compute $1212_3 + 2121_3$ and $1212_3 \times 2121_3$ in trinary format.
2. How many 12-digit binary numbers between 100000000000 and 111111111111 have at least 4 zeros?
3. Express $2^{(2^4)}$ in binary. Express $2^{(4^2)}$ in binary. Try the same with $2^{(2^8)}$ and $2^{(8^2)}$.

1.6   Base conversion

Conversion of decimal to binary – forward method

If you have a decimal number, say 1776, and you wish to convert it to binary, it may help to first determine the required *bits*, that is the binary digits. It seems natural to be most interested in the bits with the largest place values, since those seem to be the most important to the number. So notice first that, since $2048 > 1776$, and $2048 = 2^{11}$, we will be looking for, at longest, a binary number $b_{10}b_9b_8b_7b_6b_5b_4b_3b_2b_1b_0$ with 11 bits, so that

$$1776 = b_{10}2^{10} + b_9 2^9 + b_8 2^8 + b_7 2^7 + b_6 2^6 + b_5 2^5 + b_4 2^4 + b_3 2^3 + b_2 2^2$$
$$+ b_1 2^1 + b_0 2^0.$$

Since $1776 \geq 2^{10}$, we must have that $b_{10} = 1$. That gets the first bit, and to find the others, subtract 2^{10} and continue in the same fashion for the remainder $1776 - 1024 = 754$. In general, for the ith bit, if the number m remaining satisfies $2^{i+1} > m \geq 2^i$, we set the ith bit to one, and subtract 2^i, otherwise we set the ith bit to 0 and proceed with the same m, stopping only when the bit b_0 is determined.

Here is the precise description:

Algorithm 1.5 (Forward Method of Base Conversion). Let x_n be given, with $2^{n+1} > x_n \geq 2^n$. We want to express $x_n = b_n 2^n + b_{n-1} 2^{n-1} + \cdots b_1 2^1 + b_0 2^0$

by generating the two sequences $x_n, x_{n-1}, \ldots, x_0$ and $b_n, b_{n-1}, \ldots, b_0$ in $n+1$ steps, starting with the 0th step.

At the ith step set $b_{n-i} = 1$ and $x_{n-i-1} = x_{n-i} - 2^{n-i}$ if $x_{n-i} \geq 2^{n-i}$; and set $b_{n-i} = 0$ and $x_{n-i-1} = x_{n-i}$ if $x_{n-i} < 2^{n-i}$. ♡

The method described above is an example of an *algorithm*, at each step all possible cases are considered and the procedure indicated, with attention paid to how to start, and when the method is finished. The study of algorithms is an important part of discrete mathematics, and we will consider several algorithms in this text. If you are studying computer science, you will certainly take at least one class completely devoted to algorithms.

The algorithm starts with the knowledge of how many bits an integer x has when expressed in base 2. The key to that is in the inequality $2^{n+1} > x \geq 2^n$ which is true of any $n+1$ bit number which cannot be written with fewer bits. It follows that $n + 1 > \log_2(x) \geq n$, since $\log_2(x)$ is an increasing function. Since $\lfloor \log_2(x) \rfloor = n$ we have the following

Theorem 1.6. *The positive natural number x has $\lfloor \log_2(x) \rfloor + 1$ bits.*

The expression in Theorem 1.6 uses the popular *floor* function defined by setting $\lfloor x \rfloor = n$ if $n + 1 > x \geq n$. The *ceiling*, written $\lceil x \rceil$ is also useful, $\lceil x \rceil = n$ if $n \geq x > n - 1$. Note that for any integer n we have $\lfloor x \rfloor = \lceil x \rceil = n$.

Eq. (1.4) also uses the hugely important logarithm function in one of its important roles in discrete mathematics: *the logarithm to base b, with b an integer $b > 1$, roughly measures the length of the number in that base.* The precise formula for the number of base b digits of a non-negative integer x is

$$\lfloor \log_b(n) \rfloor + 1. \tag{1.4}$$

Conversion of decimal to binary – backward method

This method seems backwards because we start with what one considers the least important bit, the one's bit. Changing the one's bit has the least effect on the value of the number. But in binary it does govern an important aspect – whether the number is even or odd. Since 1776 is even, its 2^0 bit is 0.

In general, you can also decide evenness by dividing by 2 and checking the remainder: $1776 = 2 \cdot 888 + 0$. The next thing to notice is that, multiplying by 2 in binary, is just like multiplying by 10 in decimal: you shift all the bits one place to the left and add a zero at the end. If we knew 888 in binary, we could add a zero at the end and have the binary expression for 1776. Of course, finding the bits for 888 is the same mathematical problem, applied to a smaller number. We find that we just need to keep dividing by two, and recording the remainders. As we continue, we will be recording the bits of 1776 is reverse order:

$$1776 = 2 \cdot 888 + 0$$
$$888 = 2 \cdot 444 + 0$$

$$444 = 2 \cdot 222 + 0$$
$$222 = 2 \cdot 111 + 0$$
$$111 = 2 \cdot 55 + 1$$
$$55 = 2 \cdot 27 + 1$$
$$27 = 2 \cdot 13 + 1$$
$$13 = 2 \cdot 6 + 1$$
$$6 = 2 \cdot 3 + 0$$
$$3 = 2 \cdot 1 + 1$$
$$1 = 2 \cdot 0 + 1$$

We find that, $1776_{10} = 11011110000_2$, or $1776 = 2^{10} + 2^9 + 2^7 + 2^6 + 2^5 + 2^4$.

In general, given a number m, we convert it to binary by dividing by two, recording the remainder, and continuing the same procedure with the quotient, stopping only when the quotient is zero. The remainders record the bits of the number m in reverse order.

A procedure of this kind, which "calls itself", is referred to as *recursive*. Notice that this second method does not require us start by computing how many bits are required. That information is discovered along the way. Where? On the other hand, in compensation, we need to check a stopping condition.

Algorithm 1.7 (Backward Method of Base Conversion). Let $x_0 \geq 0$ be given. We want to express $x_0 = b_n 2^n + b_{n-1} 2^{n-1} + \cdots b_1 2^1 + b_0 2^0$.

Starting with the 0th step, generate b_0, b_1, b_2, ... and x_0, x_1, ... using division with remainder at the ith step: $x_i = 2 \cdot x_{i+1} + b_i$; stop the procedure when $x_{i+1} = 0$.   ♡

The forward method, where we know a priori what is to be done and how often to do it, is often called *iterative*, with the instructions at each step being *iterated*.

Other number bases

Converting a number to other bases is done analogously. For example, to convert 1776 to base 9 by the backward algorithm you would do

$$1776 = 9 \cdot 197 + 3$$
$$197 = 9 \cdot 21 + 8$$
$$21 = 9 \cdot 2 + 3$$
$$2 = 9 \cdot 0 + 2$$

giving $1776 = 2 \cdot 9^3 + 3 \cdot 9^2 + 8 \cdot 9^1 + 3 \cdot 9^0 = 2383_9$.

For the forward method, we start with Eq. (1.4) which tells us that we need 4 base 9 digits. Base 9 has more than two digits, so it is not enough, as with binary, to simply start by subtracting 9^3 and moving on immediately to the lower powers. You would either repeatedly subtract, as many as eight times, or divide by 9^3, noting the quotient and remainder before moving on to divide by 9^2, etc. This seems to make the backward method look more attractive since there you divide repeatedly by an unchanging smaller number, the base. Why not try writing out the general alogorithm explicity?

Exercises

1. Convert 6464 (decimal) to binary, using both methods. Show your work.
   Convert 6464 also to base 8 (octal) and base 16 (hex, or hexadecimal. Use A, B, C, D, E, F, for 'digits' 10, 11, 12, 13, 14, and 15). Use any method you like. If you think for a minute or two, you might see that there is a trick.
2. Convert 6464 to base 3 and to base 5.
3. For base 3, you don't have to use remainders 0, 1 and 2. Instead, you could take remainders 0, 1, and -1. Since -1 is ugly to look at as a digit, we'll invent a font: *one* ↑ for 1; *none* 0 for 0; and *mone* (pronounced mun) ↓, for -1.
   So one mone none mone, or ↑↓ 0 ↓, or [1][−1][0][−1], represents the number $3^3 - 3^2 - 3^0 = 17$.
   Write out the first 20 numbers in one/mone notation.
   Convert 1776 to the one/mone system.

1.7 Case study: Towers of Hanoi

If you like mathematical puzzles, or know someone who does, you probably have seen a model of the Towers of Hanoi. There are many versions, and variants. This is how I first encountered it. There are three posts, representing heaven, hell, and earth. There are seven round flat disks of various sizes which fit over the posts, representing parts of society.

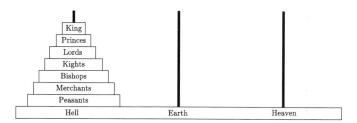

The disks start all stacked, largest to smallest, on the post representing hell, and your task is to move them, one at a time, from one post to another, so that in the end the whole tower is moved to heaven. The restriction is, that you must never place a larger disk on top of a smaller disc. So, for example, the merchant may never be placed on top of the knight.

Can the task be done in fewer than 100 moves? What about 50 moves? What is the most efficient method of solution?

As a first step, since the discs are stacked in decreasing size, and each legal move preserves that property, we know that the configuration after any number of legal moves has the discs stacked in decreasing size on the three poles. How many such legal configurations are there? It is possible to use the multiplicative principle to compute that number, and it is an upper bound on the number of moves that it takes to complete the puzzle – why?

1.8   Case study: The Binomial Theorem

We looked earlier at Pascal's Triangle, one of the most famous discrete arrangements of numbers. It is not at all surprising that those numbers are conveniently arranged in a triangle, nor that the numbers in the triangle have a mirror symmetry about the center line. But there are many curious relations in this triangle, which are harder to explain. Here is one, which I have heard called the *two-three relation*:

> *If you add up the rows of Pascal's Triangle, from left to right, with each entry scaled by an increasing power of two, they will sum to a power of three.*

Let's try:

$$\binom{4}{0}2^0 + \binom{4}{1}2^1 + \binom{4}{2}2^2 + \binom{4}{3}2^3 + \binom{4}{4}2^4$$
$$= 1 \cdot 1 + 4 \cdot 2 + 6 \cdot 4 + 4 \cdot 8 + 1 \cdot 16 = 81 = 3^4.$$

Puzzling?

Here is another curious relation. You have certainly noticed that the rows of Pascal's triangle appear when multiplying polynomials:

$$(a + b)^2 = 1a^2 + 2ab + 1b^2$$
$$(x + 1)^4 = x^4 + 4x^3 + 6x^2 + 4x + 1$$
$$(y - z)^5 = x^5 - 5x^4y^1 + 10x^3y^2 - 10x^2y^3 + 5x^4y - y^5$$

noting that the 1 on either end of the row is often not written explicitly. These two phenomena are coming from the same source, the Binomial Theorem. Binomial is not the name of a famous mathematician, it refers to a polynomial with two terms, say p and q, so written as $(p+q)$. The terms p and q may be individual letters, or something much more complicated, but regardless, the question the Binomial Theorem addresses is, what happens when we take $(p + q)$ to an integral power?

Theorem 1.8. *Let p and q be algebraic terms and let n be a positive integer. Then*

$$(p+q)^n = \sum_{k=0}^{n} \binom{n}{k} p^k q^{n-k}.$$

By algebraic terms, we are being purposely vague, intending the result to apply, to numbers, variables, expressions, etc., in fact, to any situation where addition and multiplication make sense, are commutative and the distributive law holds.

If you want to actually do the multiplication required by the integral power:

$$(p+q)^n = (p+q)(p+q)(p+q)\cdots(p+q),$$

the distributive law requires that you get one term resulting from, for each of the n terms, a choice of p versus q. If $n = 6$ and you were alternating in your choices, that would give you the term $pqpqpq = p^3 q^3$. In general, if the number of p's you choose is k, then you have the term $p^k q^{n-k}$. How many times does that occur? Once for each way to choose k factors from all n factors, that is $\binom{n}{k}$. That is why, gathering them all together in the product, we have the term $\binom{n}{k} p^k q^{n-k}$.

This proof of the Binomial Theorem is only convincing if you have a thorough understanding of how the distributive law works. We will see another proof later on. The binomial theorem explains the two-three relation completely:

$$3^n = (2+1)^n = \sum_{k=0}^{n} \binom{n}{k} 2^k 1^{n-k} = \sum_{k=0}^{n} \binom{n}{k} 2^k.$$

Here are two others:

$$2^n = (1+1)^n = \sum_{k=0}^{n} \binom{n}{k} \tag{1.5}$$

$$0 = (1-1)^n = \sum_{k=0}^{n} \binom{n}{k} (-1)^k \tag{1.6}$$

which state that the sum of the entries in any row in Pascal's Triangle is 2 to the power of that row number, and the alternating sum of the entries in any row of Pascal's Triangle must be 0. The last is obvious for odd rows by cancellation, e.g., $1 - 3 + 3 - 1 = 0$, but is certainly not obvious for the even rows like the 4th, which says, $1 - 4 + 6 - 4 + 1 = 0$. There is a non-algebraic explanation for this oddity in the next chapter.

1.9 Case study: The Guarini Problem

The Guarini Problem is a centuries old problem involving the movement of chess pieces. The starting position is four knights placed on the corners of a 3×3 section of a chessboard, with the black knights occupying the top two corners, and the white knights opposing them on the bottom two corners. The problem is to move the knights so that the black knights end on the bottom two corners and the white knights are on the top two. The knights must never leave the 3×3 area, and each move must be a legal knight move, but they do not have to take turns. So you can move the same knight several times in a row if you like.

The problem is certainly discrete. The issue is how, or whether, mathematics is involved, that is, how the problem is to be approached. If you just start moving pieces, you will discover that the problem is quite tricky, and the goal cannot be achieved in just a few moves. As the number of moves which you try increases, your choices are:

- Continue to just move at random, hoping to hit on a solution.
- Give up.
- Impose an order on your search.
- Analyze the problem.

Only with the final two can you be said to be doing discrete *mathematics*.

It is best to try the problem out for yourself before you read on. In this book we have a special symbol for this:

✠

Whenever you see it, you are encouraged to pause and think before reading on.

A favorite analysis is something like the following. There are nine squares in the 3×3 array, with 8 on the outside, forming a ring. If you mentally deform the ring, keeping the connections as they are, you can turn the 3×3 board into a circle

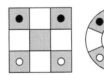

and you can ignore the color of the central region, corresponding to the middle square, since the knights never go there. They always stay on the boundary.

What does a knight's move mean in this less familiar territory? If the moves correspond to something very complicated, not matching the uniformity of the circle, then this problem transformation will not be very helpful. (But even then, the whole idea of "problem transformation" may make you think of something even more clever!)

If you are ahead of me and have already tried it, then you have discovered that knight moves on the 3 × 3 board transform very nicely indeed to rotations of the knight around the ring by 135°, 3/8 of the way around, clockwise, or counterclockwise. And that is just the hint you need. It says that the natural thing to do is not to try to reflect the knights across the line of symmetry, but to turn them around 180° degrees. As in the diagram below:

That means 4 moves for each knight, 16 altogether. That seems like a lot of moves for this little problem, but if you paused at the ✠ symbol and tried on your own, you probably gave up after many more moves than that! It is natural to start trying to move the knights directly across, since an individual knight with nothing in the way requires only two moves to cross, but the group of four get in each other's way very badly with that approach.

1.10 Case study: Red rum and murder

The fact that "murder" spelled backwards gives "red rum" was noticed long ago, and seemed compelling to those reformers over a century ago who wanted to ban alcoholic beverages. The wordplay appeared in a newspaper puzzle, arranged in a diamond grid as shown in Fig. 1.1a, with the object of the puzzle being to count all the ways to spell out the slogan, moving one letter at a time, one letter up or down, left or right.

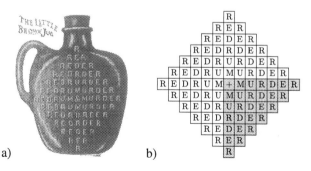

a) b)

FIGURE 1.1 a) An ancient puzzle. b) Reduction: Spelling "murder" in the (SE) quadrant.

As you start to think of how to count the ways, you quickly abandon the idea of actually listing them all. A methodical counting is clearly required. And as you work out a strategy, pay particular attention to how the puzzle might involve the multiplicative principle, and Pascal's Triangle.

Here are some insights:

- With the plus sign, the slogan redrum+murder is *palindromic*, that is, it reads the same forwards and backwards.
- You must start and end on the outer ring of R's.

This is because the inner ring of R's is separated from the letter E except across a diagonal.

- You must be in the center of the puzzle when you are half way through.

These insights give you the first reduction of the puzzle.

The only way to spell out the slogan, is to start on the outer ring of R's, move to the middle, and return.

So for each way from the exterior to the center, you can continue with any path from the center to the exterior, and these are the reverses of one another, the multiplicative principle applies. You need only count the paths from the center to the exterior, and square that.

You might ask, which is better, to count the paths from the interior to the center or from the center to the exterior? Each has its advantages. Here is an insight which is harder to see or express the other way:

- As you move from the exterior to the center, if you ever enter the long vertical expression of the slogan in the middle, or the long horizontal one, you can never leave it.

That means that, if you start in one of the four quadrants, call them (NE), (NW), (SE), and (SW), then you can never leave that quadrant. This tells us that we should use this symmetry to first focus completely on one quadrant, in one direction, and then assemble the full solution, see Fig. 1.1b.

Perhaps that is the only hint you need. We'll come back to this case study later.

1.11 Case study: Tit for tat, nim

Simple two person games are a good source of discrete problems. To play *cramcheck* you only need a checkerboard and an ample supply of checkers. The two players take turns placing one checker on the board. A player loses if there is no room for him to place a checker. Of course, the checkers are not restricted to be placed only within the marked squares – that would be no fun at all. See Fig. 1.2a. A simple strategy for the first player to win cramcheck is to initially place the first checker exactly in the middle, and thereafter to place each checker at the same distance from the center as the previously placed checker, but 180° around the board. If the first player consistently follows that strategy, each of his moves leaves the array of checkers such that it is symmetric with respect to turning the board 180°. So, using this tit-for-tat strategy, if the second player can find a place for a checker, symmetry insures that the first player can match the placement on the other side of the board in his next move, see Fig. 1.2b. Do you see why the first checker must be placed exactly in the middle?

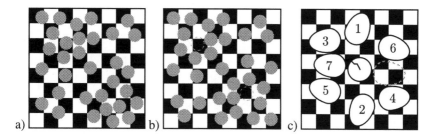

FIGURE 1.2 Cramming checkers on a board. b) and c) are symmetrically placed.

✠

It is because the center is the only location to place a single checker so that the board has rotational symmetry. Most people have a poor eye for 180° symmetry, and so it is possible for this simple strategy to go undiscovered by the opponent. On the other hand, you may find that 180° may be awkward to adhere to. Mirror symmetry is much simpler to recognize and employ, but it is possible for the second player to defeat a mirror image strategy. How?

It is said that Catherine de' Medici knew and used the 180° strategy, but was stymied when it was proposed to play the game with eggs instead of checkers. She was at a loss for a first move until she thought of jamming the first egg, hard boiled, firmly upright on its end in the middle of the board. After that, she proceeded to win, tit-for-tat.

Cramcheck has continuous and discrete aspects, but the game of *nim* is completely discrete and has a similar tit-for-tat strategy. Nim starts with a quantity of coins, distributed into piles. The two opponents take turns. Each player selects one pile, and removes from that pile as many coins as he pleases, being sure to take a least one. He may take the whole pile, but he may not alter any other pile in that turn. The winner is the one who takes the very last coin. The game may be played with large numbers of coins, say 50 coins in 10 piles of 5, or just a few, say 15 coins in piles of 1, 2, 3, 4, and 5. Why not try?

✠

If there are an even number of piles, with the numbers of coins occurring in pairs, like 22, 75, 75, and 22, then the second player can win by copying the first player, tit-for-tat. If the piles are almost balanced in pairs, like 22, 75, 100, and 22, then the first player can win by balancing the collection, that is, taking 25 coins from the pile of 100, and then following his opponent tit-for-tat.

But what if the piles are more random, like 65, 85, 19, and 21? The piles of coins seem so far out of balance that we might have to consider several layers of moves and countermoves. But, actually, the first player can "balance" them in a single move! The balancing principle is not obvious. It is hidden in the binary representations of the numbers of coins in the piles.

```
p1:    65  1000001  |  65  1000001  |  65  1000001  |  65  1000001
p2:    85  1010101  |  71  1000111  |  85  1010101  |  85  1010101
p3:    19  0010011  |  19  0010011  |   1  0000001  |  19  0010011
p4:    21  0010101  |  21  0010101  |  21  0010101  |   7  0000111
```

Given k piles of coins with p_i coins in pile i, the collection of piles is *in balance* if the number of 1's in each bit position of the binary representations of p_1, p_2, ..., p_k is even.

In the example, the four piles are out of balance because of the bits in the 2^1 and 2^3 positions. Changing the bit in the largest out-of-balance position, 2^3, from a 1 to a 0, such as for 85, 19 and 21, means decreasing the number of coins in the altered pile, even if bits in smaller positions are increased from 0 to 1.

Once the pile to be altered is chosen, the number of coins to be left in that pile follows the rule: *The bit in position 2^i is set to be 1 if the number of 1's in the 2^i position for the remaining piles is odd. It is set to be 0 otherwise.* This balances the piles, and if balancing requires the largest bit to be altered being changed from 1 to 0, the new value is smaller, and the change is made by taking away coins.

Note that an odd number of piles can be in balance!

For the example, you get to a balanced position, and are on your way to winning the game, if you take 14 coins from either the pile with 85 or the pile with 21; or if you take 18 coins from the pile with 19. (If you apply the balance rule to the pile with 65 you can balance the piles, but by *adding* 10 coins!)

If the opponent makes a move with the piles in balance, no matter which pile he chooses or how much he takes, the result leaves the piles out of balance. Then the next player can restore them to balance in the same way as before, ending with the piles being put into balance by the taking of the last coin. The player balancing the piles has won the game.

The tit-for-tat strategy for Nim is not practical without practice or without being completely comfortable with binary numbers. For good simple test cases to try with a partner, piles with three consecutive two digit numbers are recommended, like 22, 23, and 24.

1.12 Summary exercises

You should have learned about:

- Discreteness
- Mathematics
- The Multiplicative Principle
- Independence and Weak Independence
- Ordered and Unordered Selections
- Binomial Coefficients and Pascal's Triangle
- The recursive formula for Binomial Coefficients
- The Binary Number System, and other number bases

- The idea of algorithm

1. Order the following musical instruments with respect to discreteness: guitar, violin, piano, tuba, bugle, trombone, bass drum, and clarinet. Justify your choice.

2. The statement "Lightning never strikes twice in the same place" is asserting that lightning strikes are
   a) additive
   c) dependent
   b) multiplicative
   d) independent

3. A user PIN for the website www.sendmespam.com is a string of 5 digits, the 0th digit of which must be 0 and the digits in odd positions must themselves be odd.
   Can a hacker discover your PIN in one hour by trying 1 pin per second? Show your reasoning.

4. A Las Vegas slot machine called EXTAZY! has 6 tumblers, each with 6 positions corresponding to the 6 letters of "extazy". The machine pays out if all tumblers return either x, y, or z. The machine is designed so that there is always at least one x, y, or y returned among the 6 tumblers.
   How many different results can the machine return? How many different winning results can the machine return? Should you play it?
   In each case justify briefly your answer.

5. How many seven-digit numbers only contain odd digits? How many of them have 1 as the last digit? How many of them contain 7 exactly once?

6. How many n-digit numbers only contain odd digits? How many of them have 1 as the last digit? How many of them contain the digit 7 exactly once?

7. Flip a coin n times and record H if the coin lands on its head and T if it lands on its tail. How many outcomes are possible? How many possible outcomes have at least as many Hs as Ts? How many outcomes have equally as many heads and tails? Check your answers for $n = 2, 3, 5$, and 10.

8. Show algebraically that $\binom{n}{k+1}/\binom{n}{k} = (n-k)/(k+1)$. Use that to show $\binom{n}{k+1}/\binom{n}{k} \geq 1$ if $k \leq (n-1)/2$, proving Theorem 1.2.

9. a) How many anagrams of 'alligator' are there? and b) How many anagrams of 'alligator' do not have all the vowels clumped together, like 'llaiaogtr'.

10. Convert 1000 (decimal) to binary, using both methods. Show your work.
    Convert 1000 (decimal) to trinary, using both methods. Show your work.

11. Express $2^{(2^5)}$ in binary. Express $2^{(5^2)}$ in binary.

12. Find a base b for which the one's digit of the representation of 10! in base b is not 0.

Chapter 2

Basic set theory

2.1 Introduction to sets

A *set* is a collection of objects.

For instance, we may consider the set of letters used in spelling Mississippi. Let's call this set X, since one often uses capitals to denote sets. In the collection we have both letters s and i but not the letter t. We say s and i are both *elements*, or *members*, of X and write $s \in X$ to express s *is an element of X*. We also write, $\pi \notin X$, or $b \notin X$.

No other issue need be considered in the definition of a set; membership is all. For a set with very few members, it is common to simply list them all. For our initial example we may write $X = \{m, i, s, p\}$, always using "curly braces" to enclose the list of members. The letter m is the first element written on my list, but that does not mean that it is the first element of the set X, even though it is the first letter of the word "Mississippi". I could just as well list the elements in alphabetical order and still specify the same set, $X = \{i, m, p, s\}$. If I am not paying attention, or if the elements of the set are being listed by a computer which has not been programmed to remove the duplicates, the same set X could be correctly notated as $X = \{m, i, s, s, i, s, s, i, p, p, i\}$. We are not saying that these are different sets which refer to the letters in "Mississippi", we are saying that these are different ways of notating exactly the same set X:

$$X = \{m, i, s, p\} = \{i, m, p, s\} = \{m, i, s, s, i, s, s, i, p, p, i\}$$

Each expression accomplished the necessary goal of specifying which objects are elements of the set.

Every set must be *well-defined*, that is, the requirements of membership must be clear. It is easy to write *the set of words in the English language*, and we might continue to chat along merrily for quite some time as if we know what we are talking about, but that phrase is not sufficient to define a set. Is *ain't* an English word? It is in most dictionaries now, but so what. Who put dictionary editors in charge? Maybe we should ask the Poet Laureate of the United States, or the King of England, or the Supreme Court? The point is not how we could resolve the question, if the question is interesting to us. The one defining the set has the burden to making the definition clear *in the beginning*. Writing *the set of words in the English language* is not to define a bad set, or a problematic set, it is a failure – it doesn't define a set at all.

Discrete Mathematics With Logic. https://doi.org/10.1016/B978-0-44-318782-7.00007-1

Requiring sets to be well-defined means that it is actually quite difficult to define and use sets in the real world, with all its exceptional cases, and using the natural language which we speak, with all its ambiguities and shifting and evolving definitions. The clarity required to specify sets is most easily found in artificial worlds governed by abstract rules written in a rigidly controlled language – mathematics, computer science, games, and the law.

For a set to be well-defined, it is not necessary that membership be easy to determine, or even possible. The set $\mathbb{P}$ of prime numbers, integers which are at least 2 and which have no positive integer factors except themselves and 1, is clearly specified. We know that $7 \in \mathbb{P}$ and $8 \notin \mathbb{P}$, but what about $2^{(10^{10}!)} + 1$? That number is so big that it might never be practical to ever determine if it is prime or not. But it is still clear *what* we have to decide to make that determination. The set $\mathbb{P}$ is well-defined, but there are many numbers whose membership in $\mathbb{P}$ is unknown to us.

The cardinality of a finite set

For a finite set, S, the number of distinct elements in the set is called its *cardinality*, and we denote it by $|S|$. So $|\{m, i, s, p\}| = 4$, which seems obvious until you write $|\{m, i, s, s, i, s, s, i, p, p, i\}| = 4$.

Special sets

The following sets are used throughout mathematics and computer science and each has a special symbol. Most of our sets will be built from these.

Naturals The set of all *natural numbers*, the non-negative whole numbers, is denoted by $\mathbb{N} = \{0, 1, 2, 3, \ldots\}$. So 3, 0, 1776 $\in \mathbb{N}$ and -1, $3/4 \notin \mathbb{N}$. (Beware that we are taking $0 \in \mathbb{N}$. Other texts may differ.)

Integers The set of all *integers*, positive and negative whole numbers, is denoted by $\mathbb{Z} = \{\ldots, -3, -2, -1, 0, 1, 2, 3, \ldots\}$. So 3, 0, $-22^{23} \in \mathbb{Z}$, but -22^{-23}, $-\pi/4$, and ∞ are not elements of $\mathbb{Z}$.

Rationals The set of all *rational numbers*, described as integer fractions or as those decimals numbers which, after the decimal point, are finite or repeating, is denoted by $\mathbb{Q}$. So $\frac{-1}{2}, \frac{22}{7}, 0.35\overline{32} \in \mathbb{Q}$. $\sqrt{2} \notin \mathbb{Q}$.

Reals The set of all *real numbers* is denoted by $\mathbb{R}$. So π, e, γ, $1776^2 \in \mathbb{R}$. $\sqrt{-1} \notin \mathbb{R}$.

The Empty Set The empty set is the set with no elements. It is written $\{\}$ or, preferably, with the special symbol $\emptyset$. The empty set is an essential component in the theory of sets, and it must be handled and notated correctly. The empty set has cardinality zero, $|\emptyset| = 0$.

In addition to these, the following sets, primes, digits, bits, letters, will appear throughout this text.

Primes The set of all *prime numbers*, $\mathbb{P}$, natural numbers at least 2 whose only non-negative factors are themselves and 1.

Digits The set of all (decimal) *digits*, $\mathbb{D} = \{0, 1, 2, 3, 4, 5, 6, 7, 8, 9\}$.

Bits The set of all (binary) *bits*, $\mathbb{B} = \{0, 1\}$.

Alphabet The set of all 26 characters in the Roman *alphabet*, $\mathbb{A} = \{a, b, c, d, e, \ldots, x, y, z\}$.

Bracket notation

In the previous section the listing notation started to fail us. A list which relies on ellipses, the . . ., cannot really be said to properly specify its contents. Part of this difficulty we will resolve later, when we specify sets recursively. In general, a more valuable notation for an infinite set, or a large finite set, is the *bracket notation*. One specifies first either the notation for the elements to be specified, or a known set where the elements are to be taken from, and ends with a list of properties which determine the membership in the set to be defined. Separating the two parts is the | symbol, which is to be read as "such that". So the notation looks like this:

$$\{x \in A \mid \text{property 1; property 2; } \ldots\}$$

or

$$\{x \mid \text{property 1; property 2; } \ldots\}.$$

So $\{n \in \mathbb{Z} \mid n < 10; n \geq 0\}$ describes precisely the set $\mathbb{D}$.

Using this notation, we can specify the rationals in terms of the integers by writing

$$\mathbb{Q} = \{p/q \mid p \in \mathbb{Z}; q \in \mathbb{N}; q \neq 0\}.$$

If instead we write

$$\mathbb{Q} = \{p/q \mid p \in \mathbb{Z}; q \in \mathbb{Z}; q \neq 0\}$$

then we get exactly the same set, but with more duplicates, more elements of the set satisfying the conditions of membership in different ways.

Exercises

1. Which of the following sets are, in your opinion, well-defined? Justify your response in a word or two.
   — The set of football teams
   — The set of ping pong balls
   — The set of hairs on your head
   — The set of laws in the USA
2. Which of the following sets are, in your opinion, well-defined? Justify your response in a word or two.

    a. The set of integers which are products of two primes.
    b. The set of game positions in chess.
    c. The set of all real numbers whose absolute value is greater than -1.
    d. The set of integers which are useful in the study of the game Go.
    e. The set of integers which, when spoken aloud, cause a genie to appear and grant you three wishes.
3. Write each of the following in set notation, without words if possible: (Your answers will vary.)
    a. A, the set of odd integers
    b. B, the set of real numbers whose square is an odd integer
    c. C, the set of rational numbers which are an integer power of π
    d. D, the set of real numbers with a decimal representation using only digits 5 and 6

2.2 The power set

Subsets

We say that A is a *subset* of B, and write $A \subseteq B$ if every element $x \in A$ is also an element of B. In terms of our special sets we have $\mathbb{P} \subseteq \mathbb{N} \subseteq \mathbb{Z} \subseteq \mathbb{Q} \subseteq \mathbb{R}$.

Notice that it is *not* true that $\mathbb{P} \in \mathbb{N} \in \mathbb{Z} \in \mathbb{Q} \in \mathbb{R}$. If we say $\mathbb{P}$ is in $\mathbb{N}$, or $\mathbb{N}$ contains $\mathbb{P}$ we are making a vague statement which is true or false depending on whether containment is to be understood in terms of being a subset or an element of the set in question. Many problems are avoided if we discipline ourselves to use the words specifically chosen just for set theory, "an element of" or "a subset of". If you want to talk or write more informally, try to always ground things with the precise unambiguous mathematical notation: "The digits are in the rationals, $\mathbb{D} \subseteq \mathbb{Q}$".

There is also a notion of a *proper subset*, where $A \subseteq B$ but $A \neq B$, and this is often notated as $A \subset B$, but some texts use $\subset$ for $\subseteq$, so there is some ambiguity. We will not use the $\subset$ symbol in this text.

Showing $A \subseteq B$

To show that $A \subseteq B$, one must check every single element of A for membership in the set B. This is fine for very small sets, but is not generally practical. What is more effective is, if A is given by properties, you consider a general element of A, ("Let $a \in A$"); then note the properties which a must satisfy because of its membership in A; then show that, because of those properties, a satisfies the properties required for membership in B; allowing you to finally conclude that $a \in B$.

So, for instance, let's establish that $\mathbb{Z} \subseteq \mathbb{Q}$. Recall that $\mathbb{Q} = \{p/q \mid p \in \mathbb{Z}; q \in \mathbb{N}; q \neq 0\}$. Let $k \in \mathbb{Z}$. We also have $1 \in \mathbb{N}$, and since $k = k/1$, and $1 \neq 0$, we have $k \in \mathbb{Q}$, as required.

Definition 2.1. The *power set* of a set A, denoted by $\mathcal{P}(A)$, is the set of all subsets of A, $\mathcal{P}(A) = \{X \mid X \subseteq A\}$. ♠

So the elements of $\mathcal{P}(A)$ are themselves sets. This is not an unusual situation, or a weird special case. It often arises, and makes even more essential that you are careful about using the correct vocabulary for the different notions of containment.

Example 2.2. If X is a finite set, say $X = \{i, m, p, s\}$, then X has only a finite number of subsets and we can easily list them all. Of course $X \subseteq X$. Don't let the "sub" in subset mislead you. Equality is clearly allowed by the definition. The subsets with smaller cardinality are obtained by removing elements of X; cardinality 3, $\{i, m, p\}$, $\{i, m, s\}$, $\{i, p, s\}$, and $\{m, p, s\}$; cardinality 2, $\{i, m\}$, $\{i, p\}$, $\{i, s\}$, $\{m, p\}$ $\{m, s\}$ and $\{p, s\}$; cardinality 1, $\{i\}$, $\{m\}$, $\{p\}$, and $\{s\}$; and finally we are forced to consider the set with all elements removed, the *empty set*, $\{\}$.

$$\mathcal{P}(X) = \{\{i, m, p, s\}, \{i, m, p\}, \{i, m, s\}, \{i, p, s\}, \{m, p, s\},$$
$$\{i, m\}, \{i, p\}, \{i, s\}, \{m, p\}, \{m, s\}, \{p, s\}, \{i\}, \{m\}, \{p\}, \{s\}, \{\}\}.$$

The last subset considered, $\{\}$, cannot be omitted. In total $|\mathcal{P}(X)| = 16$. ◇

For any set A, it is true that $\emptyset \subseteq A$, since $\emptyset$ contains no elements to violate the subset condition. As we saw in the previous example, it can also happen that the empty set is an element of a set. So $\emptyset \notin \{i, s, m, p\}$, but $\emptyset \in \mathcal{P}(\{i, s, m, p\})$.

The set $\mathcal{P}(\{i, s, m, p\})$ has sixteen distinct elements, and one of them is $\emptyset$, so $\emptyset \in \mathcal{P}(X)$, $\{\emptyset\} \subseteq \mathcal{P}(X)$ and, look carefully at this, $|\{\emptyset\}| = 1$. The empty set is not "nothing". It is a set, so the fact that it is an element of $\{\emptyset\}$ counts. In particular, $\emptyset \neq \{\emptyset\}$!

Definition 2.3. The set of *k-subsets* of a set A is defined by

$$\mathcal{P}_k(A) = \{Y \in \mathcal{P}(A) \mid |Y| = k\}. ♠$$

For example, $\mathcal{P}_2(\{i, s, m, p\}) = \{\{i, m\}, \{i, p\}, \{i, s\}, \{m, p\}, \{m, s\}, \{p, s\}\}$, and the set $\mathcal{P}_k(A)$ will be empty if $|A| < k$. What is $\mathcal{P}_k(A)$ for $k = 0$?

Cardinality of power sets

By the multiplicative principle, if A is a finite set then

$$|\mathcal{P}(A)| = 2^{|A|}$$

since you can choose independently for each element of A whether or not it is in the subset to be considered, and membership is the defining quality for any set. Notice that choosing *no* for each element A is counted as a legal sequence of choices, and that would be specifying the empty set.

To compute the cardinality of the set of k-subsets, $|\mathcal{P}_k(A)|$, with $k \leq |A|$, we can also use the multiplicative principle. In fact, we have already done so, since those subsets may be regarded as being obtained by an unordered selection of k objects from $|A|$ objects;

$$|\mathcal{P}_k(A)| = \binom{|A|}{k}.$$

In fact, this gives us a much better insight into many problems than the idea of *unordered selection*. We don't have to imagine artificially *choosing* the elements of the subset. They are just there in the subset.

We can use the formula for the cardinality of the power set to predict the number of elements in the sets $\mathcal{P}(\emptyset)$, $\mathcal{P}(\mathcal{P}(\emptyset))$, $\mathcal{P}(\mathcal{P}(\mathcal{P}(\emptyset)))$, etc. We have already noted that $|\mathcal{P}(\emptyset)| = 1 = 2^{|\emptyset|}$, so it is consistent with that formula. Therefore we must have $|\mathcal{P}(\mathcal{P}(\emptyset))| = 2^1 = 2$. What are the two elements? We know that the whole set, and the empty set must be subsets of every set, and for $\emptyset$ they are the same set. For $\{\emptyset\}$ they are different: $\mathcal{P}(\mathcal{P}(\emptyset)) = \{\{\emptyset\}, \emptyset\}$. Continuing, $|\mathcal{P}(\mathcal{P}(\mathcal{P}(\emptyset)))| = 2^2 = 4$. Try to write all four of them down. Be very careful of the notation. (You will see that $\{\emptyset\}$ and $\{\{\emptyset\}\}$ are also different!)

Nothing prevents us from going further, and finding $|\mathcal{P}(\mathcal{P}(\mathcal{P}(\mathcal{P}(\emptyset))))| = 2^4 = 16$, and then $|\mathcal{P}(\mathcal{P}(\mathcal{P}(\mathcal{P}(\mathcal{P}(\emptyset)))))| = 2^{16} = 65{,}536$. You can stop there, because the power set of that set has $2^{65{,}536}$ elements, which I learned in school was more than the number of atoms in the whole universe, but the universe might have gotten bigger since then. We can hardly avoid being awestruck at the progression, creating an increasingly complex variety of concepts from just the contemplation of the empty set. This mathematical "big bang" was attention getting when first discovered for the same reason that big bang is compelling in physics. Mathematicians, logicians, and philosophers contemplated being able to generate all of the intellectual universe from *nothing*. (It's not so easy!)

Exercises

1. a) List all the elements in $\mathcal{P}(\{a, b, c, d\})$.

b) List all the elements in $\mathcal{P}_2(\{a, b, c, d, e\})$.

2. Find three distinct elements in each of the following sets: You must use correct notation.

- $\mathcal{P}(\mathcal{P}(\{a, b\}))$
- $\mathcal{P}_2(\mathcal{P}(\mathbb{Z}))$
- $\mathcal{P}_3(\mathcal{P}_2(\mathcal{P}_1(\mathbb{Z})))$
- $\mathcal{P}(\mathcal{P}(\mathcal{P}(\emptyset)))$

3. Show that $\mathcal{P}_2(\{1, 2, 3, 4\}) \subseteq \mathcal{P}(\{1, 2, 3, 4, 5, 6\})$.

Is $\mathcal{P}_2(\{1, 2, 3, 4\}) \subseteq \{1, 2, 3, 4, 5, 6\}$?

2.3 Set operations

In this section we consider how to create new sets from known ones. You can regard the powerset, which takes a known set A and produces a new set $\mathcal{P}(A)$, as defining an operation on sets. Using that one operation and our basic sets $\mathbb{D}$, $\mathbb{A}$, etc., we have plenty of well-defined sets to study. But we need more.

In this section we describe four other essential operations, the union, the intersection, the complement, and the Cartesian product. (There are lots more.)

Definition 2.4. The union of two sets A and B, notated $A \cup B$, is defined by $A \cup B = \{x \mid x \in A \text{ or } x \in B\}$. ♠

It is important to understand that the "or" used in the bracket notation of the definition is meant in the scientific default, in other words, one, or the other, or both. This is the *inclusive or*, which is contrasted with the *exclusive or* which one hears in the restaurant when you are asked if you would like the vegetarian *or* the non-vegetarian entré. In this text, specifically, every "or" is presumed to be meant inclusively.

Example 2.5. Let $A = \{1, 0, a\}$ and $B = \{a, b, c, 0, \pi\}$. Then $A \cup B = \{1, 0, a, a, b, c, 0, \pi\} = \{1, 0, a, b, c, \pi\}$, where it is equally valid to write the membership list with or without duplicates. ◇

The union operation is both *commutative* and *associative*; so $A \cup B = B \cup A$, and $(A \cup B) \cup C = A \cup (B \cup C)$. If A and B are finite, then the cardinality of the union satisfies $\max(|A|, |B|) \leq |A \cup B| \leq |A| + |B|$.

The intersection is a similar construction, with "or" replaced by "and". That is good since "and" seems less ambiguous in natural language.

Definition 2.6. The intersection of two sets A and B, notated $A \cap B$, is defined by $A \cap B = \{x \mid x \in A \text{ and } x \in B\}$. ♠

Example 2.7. With the same example $A = \{1, 0, a\}$, and $B = \{a, b, c, 0, \pi\}$, it is not so easy to simply write down the intersection. It requires work. I must check each candidate element of A for membership in the set B before it can be placed in the intersection. $A \cap B = \{0, a\}$ records the result. ◇

The intersection operation is also both *commutative* and *associative*; $A \cap B = B \cap A$, $(A \cap B) \cap C = A \cap (B \cap C)$. If A and B are finite, then the cardinality of the intersection satisfies $0 \leq |A \cap B| \leq \min(|A|, |B|)$.

So, as a general idea, both the union and the intersection are obviously useful and easy to describe – the union consisting of those elements which are in A and B, and the intersection being made up of those elements which are in A and B. But before you go reading merrily along, read that last sentence again. I hope it made perfect sense to you when you read it, and it seems correct to me now as I proofread it, but do you notice that almost exactly the same words were used to describe the two different situations? It made sense because, in each case, you knew what was meant, and chose automatically which of the meanings to take, probably not even noticing the ambiguity. And that ambiguity was not in the sketchy word "or", but in our good friend "and". This is normal. It happens all the time if we communicate in natural language, which we must if we want to work on problems which occur in the natural world. One of the purposes of the use of set theory is to give us an unambiguous mathematical vocabulary to avoid confusion. You lose all the advantage if you always take the precise set theoretic formulation and re-express it back in familiar natural language and work with that. You will end up just fooling yourself. You want to gain a facility with the proper use of this new vocabulary, so don't try to avoid it.

Definition 2.8. The *complement* of the set A, denoted by A^c, is defined by $A^c = \{x \mid x \notin A\}$. ♠

The complement is a *unary* operation, like the power set, in that it acts on a single set to create a new set. The union and intersection are both called *binary* operations. The complement has a few issues in its definition and requires some care, and the notation for the complement has not reached the same degree of standardization as the union and intersection. Another popular notation is $\bar{A}$, but there are other important concepts competing for the overbar.

Example 2.9. Let $A = \{1, 0, a\}$, and $B = \{a, b, c, 0, \pi\}$ again. Clearly we have $\pi \in A^c$, and $a \notin B^c$. But we have some curious things happening here, too. We have $\emptyset \in B^c$, $\{\emptyset, \{\emptyset\}\} \in B^c$, $\sqrt{-1} \in A^c$, and $\mathbb{P} \subseteq A^c$. Yes, these are truly gigantic sets, which they have to be. For A^c, the query as to membership will be answered almost always "yes" except for the three queries associated with the number 1, the number 0 and the letter a, where the answer as to membership in A^c is "no". ◇

The situation of the previous example suits many situations which mathematicians encounter. But, practically, the complement is often encountered in the context of a particular collection of elements outside of which we have no interest. You may have restricted your attention to words in a computer program which you are writing, or characters in a game you are developing. This particular set is called the *universe*, and usually denoted by $\mathcal{U}$. If you are working with a universal set, then all elements from all sets you consider while working in that universe are drawn just from that universal set, $\mathcal{U}$. You can work in a large numerical universe consisting of numbers and sets of numbers, such as $\mathbb{R} \cup \mathcal{P}(\mathbb{R}) \cup \mathcal{P}(\mathcal{P}(\mathbb{R})) \cup \ldots$, or you may prefer to do work in a tiny universe like $\mathcal{U} = \mathbb{D} \cup \mathbb{A}$. For that tiny one, $A^c = \{x \in \mathcal{U} \mid x \notin A\}$ and consists of just 33 elements, i.e., those digits and letters not equal to 0, 1, or 1a, which for that universe is a lot. For this meaning of the complement it is also said that A^c is the complement of A *relative to* $\mathcal{U}$. If the universe is a finite set, then $|A^c| = |\mathcal{U}| - |A|$.

Definition 2.10. The *Cartesian product* of two sets A and B, notated $A \times B$, is defined by $A \times B = \{(a, b) \mid a \in A; b \in B\}$. ♠

The Cartesian product is another binary operation but, like the power set, the Cartesian product results not only in a new set of elements, but in a set whose elements are a new type of thing, *ordered pairs*, of existing elements.

Example 2.11. Again with the same example sets $A = \{1, 0, a\}$, and $B = \{a, b, c, 0, \pi\}$, in order to create $A \times B$ we have to, for each of the three elements allowed to be in the first coordinate of the pair, place any of the five allowable elements of B into the second coordinate, so $A \times B$ will have fifteen elements

$$A \times B = \{(1, a), (1, b), (1, c), (1, 0), (1, \pi), (0, a), (0, b), (0, c), (0, 0), (0, \pi),$$
$$(a, a), (a, b), (a, c), (a, 0), (a, \pi)\}.$$

Notice particularly that (a, a) and $(0, 0)$ are included because $a, 0 \in A \cap B$, but no other pair of identical coordinates is included. Notice also that, even though $(a, a) \in A \times B, a \notin A \times B.$ ◇

If A and B are finite, then $|A \times B| = |A| \cdot |B|$ by the multiplicative principle. You should observe that the Cartesian product is *not* commutative and *not* associative.

Exercises

1. Let $X = \{0, \emptyset\}$ and $Y = \{\emptyset, \pi\}$.
    a. List all elements of $\mathcal{P}_2(X \cup Y)$.
    b. List all elements of $\mathcal{P}_2(X \cap Y)$.
    c. List all elements of $\mathcal{P}_2(X \times Y)$.
2. Let $A = \{0, 2, 4, 6, 8\}$, and $B = \{1, 3, 5, 7, 9\}$.
    a. List five different elements of $A \times (B \times (A \cup B))$.
    b. How many elements does this set have? (Use the multiplicative principle)
3. Let $P = \{a, b, c\}$, $Q = \{c, d, e\}$ and $R = \{e, f, a\}$.
    a. Find all elements of $\mathcal{P}((P \cap Q) \times (Q \cap R) \times (R \cap P))$.
    b. Find all elements of $\mathcal{P}(P \cap Q) \times \mathcal{P}(Q \cap R) \times \mathcal{P}(R \cap P)$.

2.4 Set identities

Basic laws

As noted above, the union and intersection are both commutative and associative, and have a special relation with $\emptyset$. These properties can be written as identities. For all sets A, B, and C:

$$
\begin{aligned}
A \cup A &= A & A \cap A &= A \\
A \cup \emptyset &= A & A \cap \emptyset &= \emptyset \\
A \cup B &= B \cup A & A \cap B &= B \cap A \\
A \cup (B \cup C) &= (A \cup B) \cup C & A \cap (B \cap C) &= (A \cap B) \cap C \\
(A^c)^c &= A & \emptyset^c &= \mathcal{U}
\end{aligned}
$$

$$(2.1)$$

Distributive laws

Just like in ordinary algebra, where multiplication distributes over addition and gives the identity $a(b+c) = ab+ac$, the set operations also satisfy a distributive law. For sets the law works more generally in that $\cap$ distributes over $\cup$ and also $\cup$ distributes over $\cap$. So, for sets A, B and C:

$$A \cap (B \cup C) = (A \cap B) \cup (A \cap C)$$
$$A \cup (B \cap C) = (A \cup B) \cap (A \cup C)$$

De Morgan's laws

The interaction of the complement with the basic operations is expressed in De Morgan's laws. For all sets A and B:

$$(A \cap B)^c \quad = \quad A^c \cup B^c \qquad (A \cup B)^c \quad = \quad A^c \cap B^c \qquad (2.2)$$

Condition for set equality

The next result might not look like an identity now, and might not seem important enough to mention, but for us it is the most important. Some authors think of it as a key definition, but we express it as a theorem.

Theorem 2.12 (Condition for Set Equality). *Two sets A and B are equal if and only if $A \subseteq B$ and $B \subseteq A$ are both true.*

Proof. The assertion hardly merits a proof. If the sets are equal, then the subset relations hold because every set is a subset of itself. On the other hand, if each is a subset of the other, then neither can have any element which the other does not have, so the sets have the same elements, which means that they are the same set. □

The reason this obvious result comes up so often is that, generally, deciding whether two collections are exactly the same requires us to divide our attention between two sets, possibly different, at least specified in different ways, comparing all their many elements having perhaps many duplicate specifications. It is very easy to get confused. Dividing the task into two parts helps right there, but even more, showing the subset relation, instead of set equality, allows us to focus our attention on one element and its properties, rather than the whole set. Using the theorem in this way is such an essential trick that we call it the *Double Inclusion Method*. This method comes up very often when people suspect that they are looking at the same thing in two different ways. We devote the next section to the method, where we apply it to simple examples mostly based on the identities stated in this section. But keep in mind, it is Condition for Set Equality, Theorem 2.12, which is important, and it is the double inclusion method which we are trying to get practice using, not the simple identities we are applying it to.

Exercises

1. Suppose $A \subseteq B \subseteq C$. Show that $(A \times B) \subseteq (B \times C)$.
   a. Use the method we discussed of showing the subset relation.
   b. Need it be true that $(B \times C) \subseteq (A \times B)$?
2. As part of using the double inclusion method to show that $A \cup (B \cap C) = (A \cup B) \cap (A \cup C)$ we would have to show $A \cup (B \cap C) \subseteq (A \cup B) \cap (A \cup C)$ for any sets A, B and C.
   Show $A \cup (B \cap C) \subseteq (A \cup B) \cap (A \cup C)$.
3. As part of using the double inclusion method to show that $A \cup (B \cap C) = (A \cup B) \cap (A \cup C)$ we would have to show $A \cup (B \cap C) \supseteq (A \cup B) \cap (A \cup C)$ for any sets A, B and C.
   Show $(A \cup B) \cap (A \cup C) \subseteq A \cup (B \cap C)$.
   [Hint: The two cases $x \in A$, and $x \notin A$ make the argument simpler.]

2.5 Double inclusion

In this section we will be illustrating how to prove identities using the double inclusion method. We will illustrate the method three times. The first is with a less familiar identity, not found in the previous section. We will give the identity, then a proof, and then provide a long commentary on that proof.

Theorem 2.13. *Let A and B be sets. Then $\mathcal{P}(A) \cap \mathcal{P}(B) = \mathcal{P}(A \cap B)$.*

Proof. We first show $\mathcal{P}(A) \cap \mathcal{P}(B) \subseteq \mathcal{P}(A \cap B)$. Let $x \in \mathcal{P}(A) \cap \mathcal{P}(B)$. Then $x \in \mathcal{P}(A)$ and $x \in \mathcal{P}(B)$. The first fact implies $x \subseteq A$. The second fact implies $x \subseteq B$. So $x \subseteq A \cap B$, and $x \in \mathcal{P}(A \cap B)$, as required.

We next show that $\mathcal{P}(A \cap B) \subseteq \mathcal{P}(A) \cap \mathcal{P}(B)$. Let $y \in \mathcal{P}(A \cap B)$. So $y \subseteq A \cap B$. Since $y \subseteq A \cap B$, it follows that $y \subseteq A$, and as well that $y \subseteq B$. So we have that $y \in \mathcal{P}(A)$ and $y \in \mathcal{P}(B)$. Thus $y \in \mathcal{P}(A) \cap \mathcal{P}(B)$, finishing the second demonstration.

Since each set is contained in the other, the identity is true by Theorem 2.12. ☐

Commentary: The first thing to notice is "Let x". You want to show that the subset relation holds, so you have to show that each element of $\mathcal{P}(A)$ is contained in the other set. You cannot do it by checking elements one by one. You don't even know which set A is referred to. Instead, you take a general element, call it something, x, and then restrict your attention to that x, and the properties it inherits from membership in $\mathcal{P}(A)$. Then you use those properties of x in order to show that x also has the qualities required for membership in the other set. This is the power of the double inclusion method mentioned at the end of the last section; that we can focus on an individual element and its properties, instead of the two sets and their many elements.

The second thing to notice is "as required". Required by whom? If you are the proof writer, then it is you who made that requirement when you announced that you were going to show set containment, and made the first "let".

The third thing to notice is that you can probably think of many ways that you might have written this proof differently; maybe easier to follow, perhaps shorter. Maybe you don't like that I chose x for the general element when, since it turned out to be a set, it could be clearer to use X. Maybe you think there should be a little more detail, or a little less. Even in model proofs like this there is room for expression, personality, and style. How do you avoid freezing up, paralyzed by the choices? I think it helps tremendously to turn your attention from yourself to the reader, and what the reader needs to be told. All a proof is, ultimately, is a convincing argument – an argument convincing to an open minded, but sceptical reader – someone willing to be persuaded, but only after being convinced that the argument is sound and in which every detail is actually correct. That is how you should listen to proofs.

And what if you, the reader, don't agree? What then? It can happen that you see an error and you can explain the problem. But probably not. Likely as not you don't follow the argument because it is too complicated for you, or because the writer is confused and has made a mistake, or because there are differing definitions. But whatever the reason, you cannot accept the result as having been proven. That doesn't mean you can conclude the assertion is false, just that you are unconvinced by that argument.

Next, I will show you how to completely destroy the first half of the proof:

......$\mathcal{P}(A) \cap \mathcal{P}(B) \subseteq \mathcal{P}(A \cap B)$. $x \in \mathcal{P}(A) \cap \mathcal{P}(B)$, $x \in \mathcal{P}(A)$, $x \in \mathcal{P}(B)$, $x \subseteq A$, $x \subseteq B$, $x \subseteq A \cap B$, $x \in \mathcal{P}(A \cap B)$ ....

This is by far the most common student mistake. The idea seems to be that a proof gets more mathematical, and shorter, and truer, if you just leave all those non-mathematical words out. But if you leave the words out, you leave the whole argument out! Look at all those bare equations, some of which were goals to be done, some just assumptions for the sake of argument, some conclusions. They are now just asserted altogether as bare facts. As a proof, it must be wrong. You cannot convince without an argument. So you might as well try putting in the words and risk making a poor argument, rather than leaving them out and be sure that it fails.

Another point: that jumbled mess of equations in the previous paragraph does look an awful lot like what you see on the black board, or whiteboard, or slides of professionals, professors and other experts. Don't be fooled. For a presentation, the presenter is speaking, weaving those ingredients together to present a convincing argument. The slides alone are often a woefully inadequate substitute.

One last word of commentary. As you write a proof, your goal is to convince the reader. Who is this reader? It is much easier to write anything if you can imagine to whom you are writing. It is a huge mistake to tailor any argument to convince a person who you believe knows much more than you do. That is a hopeless task. So don't imagine that you are writing to your professor, or the smartest person in the room. Don't imagine, contrariwise, that your reader knows nothing. The reader who knows nothing is not interested in your proof. In general, you should imagine the reader as one of your colleagues, someone who is about at your level but who just doesn't know about this particular thing which you want to explain, say your own self as you were last week.

Now let's prove one of the set identities from the previous section by the double inclusion method,

$$A \cap (B \cup C) = (A \cap B) \cup (A \cap C)$$

and then there will be a very short commentary.

Proof. First we will show that $A \cap (B \cup C) \subseteq (A \cap B) \cup (A \cap C)$. Let $d \in A \cap (B \cup C)$, so $d \in A$ and $d \in B \cup C$. From the second fact, we are led to consider two cases:

Case 1: $d \in B$. Here, since $d \in A$, we have $d \in A \cap B$, so $d \in (A \cap B) \cup (A \cap C)$.

Case 2: $d \in C$. Now $d \in A$ implies $d \in A \cap C$, so again $d \in (A \cap B) \cup (A \cap C)$.

In either case, $d \in (A \cap B) \cup (A \cap C)$, as required.

Next we want to check $(A \cap B) \cup (A \cap C) \subseteq A \cap (B \cup C)$ and assume $e \in (A \cap B) \cup (A \cap C)$. So we consider on the one hand that $e \in (A \cap B)$, in which case $e \in A$ and $e \in B$ hence $e \in B \cup C$ and conclude $e \in A \cap (B \cup C)$; and on the other side you might have $e \in (A \cap B)$, and get that both $e \in A$ and $e \in B$ leading me to the same ending. Therefore $(A \cap B) \cup (A \cap C) \subseteq A \cap (B \cup C)$ and the identity is proved. $\square$

Commentary: The two halves are written in different styles, as if in a team with several authors. The first part is crisply laid out into cases. The second part is mostly one long run-on sentence with shifting voice, inconsistent punctuation, and other errors. As writing it may be horribly ugly, but it is valid, and that is the only mathematically important issue. You may be wondering about the 'third case', when d is in both sets – then both cases apply, so the conclusion certainly holds, and you would be wasting your time to add a third case to handle what has already been done twice, even though it is valid to do so.

Our third proof moves us just a bit beyond the set identities which we are using for practice. X will be the set of all even natural numbers, and Y and Z those numbers divisible by 3 and 6.

Theorem 2.14. *Let* $X = \{x \mid x = 2i; i \in \mathbb{N}\}$, $Y = \{y \mid y = 3j; j \in \mathbb{N}\}$, *and* $Z = \{z \mid z = 6k; k \in \mathbb{N}\}$. *Then* $Z = X \cap Y$.

Proof. First let $z \in Z$. Then $z = 6k$ for some $k \in \mathbb{N}$, and $6k = 2(3k) = 3(2k)$. Setting $i = 3k \in \mathbb{N}$ implies $x = 2i \in X$, and setting $j = 2k \in \mathbb{N}$ implies $x = 3j \in Y$. Thus $z \in X \cap Y$ as required.

Now let $z \in X \cap Y$. So $z = 2i = 3j$. Since $z = 2i$, z is even. Since $z = 3j$, j must be even. So $j = 2k$ for some $k \in \mathbb{N}$, and $z = 3(2k) = 6k$. This says $z \in Z$, as required. □

Commentary: Many people don't like that the letter z from the first half was recycled in the second half for a different number. But one finds that in double inclusion proofs quite often.

Exercises

Assertion: Let A be a set. Then $A = \mathcal{P}(A)$.

Proof: Part 1: Clearly A is in $\mathcal{P}(A)$, so $A \subseteq \mathcal{P}(A)$.

Part 2: Now to show $\mathcal{P}(A) \subseteq A$. Let X be contained in $\mathcal{P}(A)$. Then every element of X is contained in A, so X is contained in A. Since everything in $\mathcal{P}(A)$ is contained in A, $\mathcal{P}(A) \subseteq A$.

Part 3: Since $\mathcal{P}(A) \subseteq A$ and $A \subseteq \mathcal{P}(A)$, we have $\mathcal{P}(A) = A$.

1. Find any flaws in Part 1 that you can.
2. Find any flaws in Part 2 that you can.
3. Find any flaws in Part 3 that you can.

2.6 Russell's paradox

Recall when defining the complement we talked about a universal set. It is common to work with a universal set containing all the elements which you are interested in, a universal set tailored to a problem or collection of problems. The only difficulty with that is that some of the standard operations, like $\mathcal{P}$ and $\times$, create sets with elements of new types, and these elements might be "outside the universe", like in a science fiction movie. One way to avoid this is to try to define a really gigantic universe from which there is no escape. A good place to start might be

$$S \quad \text{— The set of all sets}$$

Once we have the set of all sets, we can easily construct a universal set big enough to suit to anybody anywhere by

$$\mathcal{U} = \{x \in S \mid S \in \mathcal{S}\}.$$

But $\mathcal{S}$ has a property not seen in any of the sets we have considered so far. The set $\mathcal{S}$ is an element of itself: $\mathcal{S} \in \mathcal{S}$. This looks familiar, but isn't. It is the relation $A \subseteq A$ which is true for all sets; every set is a subset of itself. What about power sets? For any of the sets we have considered, $X \in \mathcal{P}(X)$, but we have never had $\mathcal{P}(X) \in \mathcal{P}(X)$, instead having $\mathcal{P}(X) \in \mathcal{P}(\mathcal{P}(X))$.

Let's try to build a small example, a set containing just itself and the number 1. So $1 \in X$ and $X \in X$, which gives $X = \{1, X\}$. That looks ok at first, but then that says $X = \{1, \{1, X\}\} = \{1, \{1, \{1, \{1, \ldots\}\}\}\}$ and it is hard to figure out what the … even means. After all, what *is* that other element? We seem to be stuck. These guys seem not so easy to construct.

Suppose you are a conservative person, and this property makes you uncomfortable, and you don't want to consider such sets. You might want a restrict yourself to sets which are more ordinary:

$$\mathcal{R} = \{X \in S \mid X \notin X\}.$$

$\mathcal{R}$ looks much safer than $\mathcal{S}$. $\mathcal{R}$ contains precisely those sets which are not elements of themselves.

But what about $\mathcal{R}$ itself, is $\mathcal{R} \in \mathcal{R}$? On the one hand if $\mathcal{R} \in \mathcal{R}$, then $\mathcal{R}$ fails to satisfy the property specified in its definition for membership, so we would have to conclude, contradictorily, that $\mathcal{R} \notin \mathcal{R}$. On the other hand, if you surrender and accept that $\mathcal{R} \notin \mathcal{R}$, then $\mathcal{R}$ passes the requirement for being an element of the set $\mathcal{R}$, and we are confronted again with the reverse, $\mathcal{R} \in \mathcal{R}$. So we have arrived at an impasse, a contradiction. Normally that is ok. Arriving at a contradiction just means that some assumption which we have made is incorrect. Identifying the false assumption is progress. But here we don't seem to have assumed anything at all. We were just getting started constructing things.

If this makes you scratch your head, you are not alone. This conundrum was a bucket of cold water in the face of many people when Bertrand Russell expounded it a century ago, and it is called *Russell's paradox*. The problem was quickly identified as resulting from the fact that the set $\mathcal{S}$ refers to itself in its own definition; a trick used in other linguistic problems long known to logicians and philosophers. That is what we encountered with $\{1, \{1, \{1, \{1, \{1, ????\}\}\}\}\}$.

There are essentially three ways out of this problem. One is to avoid self-reference by forcing sets to be in a hierarchy, in other words, elements belong to sets, sets belong to *classes*, classes belong to *super-classes*, etc. The second way out is to make set theory *axiomatic*, as is done with Euclidean geometry, where you seem to be proving theorems about points and lines in space, but you are actually proving things about "points", and "lines", in a closed system of assumptions. We will follow the third way, which is what is done by most people who simply want to use sets to solve practical problems – which is to start with the elements we are interested in, numbers, sets of numbers, letters, game characters, words in a computer program, etc.; then expand this working universe as necessary, and not even attempt to contemplate a whole and complete system.

So the moral of this very abstract story is, beware of definitions which refer to themselves and, the universe is actually important.

Exercises

1. Overheard conversation between Pat and Mike:
   Pat: Mike, you always lie.
   Mike: Yeah, and I am lying right now.
   Who is telling the truth?
2. Sign at the bus stop:

   > The sentence below is true.
   > The sentence above is false.

   Which of these lines is true?
3. Which of these three problems involve self-reference?

2.7   Case study: Polyhedra

If you have ever played a game needing exotic dice, or investigated three dimensional symmetric objects, then you have seen the five *Platonic solids*,

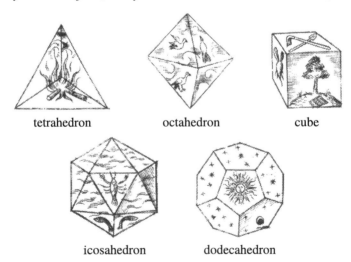

tetrahedron   octahedron   cube

icosahedron   dodecahedron

perhaps better rendered than in this hand drawing by Johannes Kepler from back in the day when four elements, earth, water, air and fire, seemed plenty. Otherwise, all but one of these interesting objects, the common cube, are probably unfamiliar to you.

One of the reasons people have been fascinated by these five objects is the fact that there are only five. There may be an infinite number of interesting three dimensional solids, but these are the only ones all of whose *hedra*, faces, are congruent regular polygons, with the same number of polygons coming together at each vertex.

In general, polyhedra may be thought of as being obtained by starting with all of continuous three dimensional space, and methodically cutting away chunks of it, leaving a solid with a finite number of flat *faces* on its surface where the cuts were made. Of course, most polyhedra will not be regular or symmetric, but in this case study we will only look at symmetric examples.

How do we make this idea of cutting away space precise? With set theory. Instead of a knife, we use a set called a *half-plane* and instead cutting, we intersect. A plane does "cut through space" with points on either side. Let $\mathbf{v}$ be a point not at the origin, and let $\mathcal{P}_\mathbf{v}$ denote the plane through $\mathbf{v}$ and perpendicular, or orthogonal, to the segment from the origin to $\mathbf{v}$. If you know some vector algebra, $\mathcal{P}_\mathbf{v} = \{\mathbf{x} \in \mathbb{R}^3 \mid \mathbf{x} \cdot \mathbf{v} = \mathbf{v} \cdot \mathbf{v}\}$, and the half-plane which $\mathcal{P}_\mathbf{v}$ bounds and which contains the origin is $\mathcal{H}_\mathbf{v} = \{\mathbf{x} \in \mathbb{R}^3 \mid \mathbf{x} \cdot \mathbf{v} \leq \mathbf{v} \cdot \mathbf{v}\}$. The cube can now be expressed as the set $C = \mathcal{H}_\mathbf{i} \cap \mathcal{H}_\mathbf{j} \cap \mathcal{H}_\mathbf{k} \cap \mathcal{H}_{-\mathbf{i}} \cap \mathcal{H}_{-\mathbf{j}} \cap \mathcal{H}_{-\mathbf{k}}$. This is more than a mere description. The set theoretic interplay of six cutting planes of C holds the key to C's "cubyness". The top square face of the cube is $f_\mathbf{k} = \mathcal{P}_\mathbf{k} \cap C$, just one of the six faces of $\mathbb{C}$.

The surface of the polyhedron, besides the finite set of *faces*, $\mathcal{F}$, also has a finite set of *edges*, $\mathcal{E}$, line segments obtained from the intersection of two *incident* faces. For example, in C, find $f_\mathbf{k} \cap f_{-\mathbf{j}}$. There is also a finite set of *vertices*, $\mathcal{V}$, whose elements are singleton sets of the form $v = \{(x, y, z)\}$ corresponding to isolated points where three or more faces have an non-empty intersection, see Fig. 2.1, where faces f_1 and f_2 intersect in edge e, and $f_1 \cap f_2 \cap f_3 = v$.

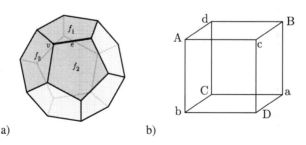

a)   b)

FIGURE 2.1   a) A dodecahedron and b) a cube with labeled vertices.

Given a set of labels of one type of object, set operations allow us to generate natural labels for the others. For example, for the cube, take the eight vertex labels to be $\mathcal{V} = \{a, b, c, d, A, B, C, D\}$, always choosing the capital and lower case version of the same letter to be opposite, and never labeling the endpoint of any edge with letters of the same case.

Once the vertex labels are chosen, it is natural to label the elements of the edge set $\mathcal{E}$ from the set $\mathcal{P}_2(\mathcal{V})$, labeling each edge with the pairs of vertex labels for the endpoints of that edge. For our cube, the elements of the set $\mathcal{E}$ consist of pairs, one upper case and one lower case, never belonging to the same letter. The twelve edges have labels:

$$\mathcal{E} = \{\{A, b\}, \{A, c\}, \{A, d\}, \{B, a\}, \{B, c\}, \{B, d\},$$
$$\{C, a\}, \{C, b\}, \{C, d\}, \{D, a\}, \{D, b\}, \{D, c\}\}.$$

Following this system, the face labels of a polyhedron will naturally be elements of $\mathcal{P}(\mathcal{V})$, with each face labeled by the subset of vertex labels corresponding to the set of vertices it contains. Since the polygon with the fewest vertices is a triangle, every face label will be a subset of $\mathcal{V}$ of cardinality at least three, and no face label will also be an edge label. For the cube all faces are squares, hence the face labels are elements of $\mathcal{P}_4(\mathcal{V})$ and consist of all $\binom{4}{2}$ pairs of the four capitals $\{A, B, C, D\}$ together with the complementary pair of lower case letters. So $\mathcal{F}$ consists of

$$\{\{A, B, c, d\}, \{A, C, b, d\}, \{A, D, b, c\}, \{B, C, a, d\}, \{B, D, a, c\}, \{C, D, a, b\}\}.$$

The tetrahedron of Fig. 2.2a with vertex set $\mathcal{V} = \{A, B, C, D\}$ is an extreme case in which the edge set is labeled by all $\binom{4}{2}$ elements of $\mathcal{P}_2(\mathcal{V})$,

$$\{\{A, B\}, \{A, C\}, \{A, D\}, \{B, C\}, \{B, D\}, \{C, D\}\}.$$

and the triangular faces are labeled by all $\binom{4}{3}$ elements of $\mathcal{P}_3(\mathcal{V}) = \{\{A, B, C\}, \{A, B, D\}, \{A, C, D\}, \{B, C, D\}\}$.

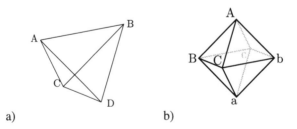

a)   b)

FIGURE 2.2   a) A tetrahedron and b) an octahedron with labeled vertices.

The elements of $\mathcal{V} = \{A, B, C, a, b, c\}$, vertices of the octahedron labeled as in Fig. 2.2b, contribute to eight triangular faces, $\mathcal{F} \subseteq \mathcal{P}_3(\mathcal{V})$. Each has a representative of each letter, and all possible choices of capital and lower case occur.

$$\{\{a, b, c\}, \{a, b, C\}, \{a, B, c\}, \{a, B, C\},$$
$$\{A, b, c\}, \{A, b, C\}, \{A, B, c\}, \{A, B, C\}\}.$$

There are $\binom{6}{2} = 15$ 2-subsets of $\mathcal{V}$, and all correspond to edges except $\{A, a\}$, $\{B, b\}$, and $\{C, c\}$.

Nice labeling choices for the icosahedron and the dodecahedron exist, but are harder to motivate and describe.

No matter how their constituent parts are labeled, the five platonic solids all satisfy

$$|\mathcal{V}| - |\mathcal{E}| + |\mathcal{F}| = 2,$$

which turns out to be true not only for the five platonic solids, but all three-dimensional polyhedra (cf. Theorem 9.14).

2.8   Case study: The missing region problem

Related to the problem of describing polyhedra is the problem of dissecting the plane into regions. This is an important, well-studied problem. Here is one version.

Consider a circle C in the plane and a finite subset $V \subseteq C$. Let D, for "disc", be the union of C with the set of all points in the interior of C. For each element of $\mathcal{P}_2(V)$, draw the line segment joining them, their chord. You now have a circle with $\binom{|V|}{2}$ chords. Into how many regions is D divided by these chords?

The answer should depend on the cardinality of V as well as how the points are placed on the circle. For some preliminary data look at Fig. 2.3. If n is even and the elements of the set V are equally spaced, then there are $|V|/2$ diameters meeting at the center of the circle, as in the case of the square and the hexagon in Fig. 2.3, and the number of regions are divisible by 4 and 6 respectively. It is mildly surprising that the fivefold symmetry of the pentagon/pentagram yields a number of regions, 16, which is not divisible by 5, but that is because of the exceptional central five-fold symmetric region. Other than that, the regions do come in three symmetric subsets containing 5 elements each, giving $3 \cdot 5 + 1 = 16$ regions all together.

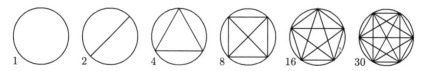

FIGURE 2.3   Dividing the circle into regions.

It seems no more than a coincidence that the number of regions for 5 points is a power of 2, until one notices that the same is true for all the examples in Fig. 2.3 except the last one. In fact, the first five cases all follow the simple formula $2^{|V|-1}$ for the number of regions. Unfortunately, it is easy to count that the last one has only 30 regions, not 32, which would be the next power of 2.

The last diagram does have that special point, the point in the very center, where three diameters meet. Perhaps we have that special point at the cost of some extra regions. Perhaps instead of looking at very symmetric elements of $\mathcal{P}_6(C)$, we should look in the other direction, asymmetric, generic, or random subsets of cardinality 6 of the circle. If the set of regular points on the circle C are perturbed a little, as in Fig. 2.4, none of the other counts are changed, but the last one does get yet one more region.

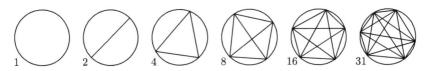

FIGURE 2.4 Dividing the circle into uglier regions.

The Missing Region Problem asks: *Can we perturb again so that the missing 32nd region appears?*

2.9 Case study: Soma

Soma is a dissection puzzle invented by Piet Hein, patented in the 1930's as a 3D puzzle. It has been in production ever since. Soma is built from 27 little cubes glued permanently into seven distinct irregular pieces. Since $3^3 = 27$, it is conceivable that the 7 irregular pieces can be assembled into a $3 \times 3 \times 3$ cube, and that is one of the more popular objectives of the puzzle, see Fig. 2.6.

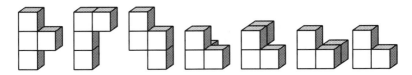

FIGURE 2.5 Non-convex assemblies of four or fewer cubes.

FIGURE 2.6 One view of an assembled Soma cube.

A reasonably attentive adult with no special mathematical skills can often solve the puzzle in less than ten minutes. It often happens, however, that another person will remain stymied by the Soma cube for hours or even days. If that happens, it is of course frustrating, but this is part of the charm of the puzzle. We will see shortly why it is that some people find themselves stumped so much longer than many others.

The method most people use is to try various combinations mostly at random. A generally successful approach of this kind is to fill in first about half the pieces, and then try to methodically vary the placement of the remainder to complete the cube. For many puzzles, committing to a partial solution and trying to complete from there would be a very bad strategy since it is common for puzzle designers to allow only for a unique solution. But Soma has 240 'solutions'.

One question is whether mathematics helps speed up the solution process. A more ambitious problem is to find and enumerate all possible solutions up to symmetry. A quick and dirty computer program could be written to examine all the combinations of possible positions of each of the pieces and check for compatibility. Unfortunately, unless one takes care, this approach yields a combinatorial explosion. You may compute that the second piece in Fig. 2.5, for instance, may be placed in the $3 \times 3 \times 3$ cube in 144 ways. A more successful approach is to apply a "branch and bound" strategy to avoid reconsidering rejected configurations for different reasons. The most sophisticated computer to have completely enumerated all possible solutions consisted of two self-programming parallel organic processors, John Horton Conway and Michael J.T. Guy, coming together in 1961, the job taking all of a Saturday to complete.

The most important observation to simplify this problem is to notice that the set S of 27 subcubes of a solution is partitioned into 4 subsets, $S = V \cup E \cup F \cup C$, having empty intersection with one another, corresponding to the different parts of the cube. The set V contains the 8 subcubes at the vertices of the large cube, the set E the 12 subcubes at the edges, the set F the 6 subcubes at the face centers, and C the one subcube at the center. It helps to imagine the subcubes of different sets in different colors. Another interesting partition is the 'checkerboard partition', $S = (V \cup F) \cup (E \cup C)$. The set $V \cup F$ has cardinality 14 and we'll imagine its elements colored red, and the set $F \cup C$ has cardinality 13 and we'll imagine its elements colored blue.

Let's consider the red/blue coloring with regard to the individual Soma pieces. There is a 4-subset of the set of pieces which must be colored two blue and two red in any solution, see Fig. 2.7. That leaves 6 red and 5 blue to be accounted for with the three elements of the complement, the subset of biased pieces. Two elements of that subset, look in Fig. 2.5 for which ones, must have either three red and one blue, or the reverse. They both cannot be biased in the same way, since then those two pieces would account for 6 cubes of one color all by themselves.

FIGURE 2.7   These four pieces have only balanced two-colorings.

Thus the two larger biased pieces must be biased oppositely, their bias canceling one another out. Therefore it must be true that the little piece must be biased in favor of red.

So we have found two conditions which obstruct a partial solution ever being completed. (There is another nice one you can find by considering the corners alone.) If the two larger biased pieces are placed so that their biases do not cancel, it is impossible to complete the Soma cube, just like if the smallest piece is not placed to be biased in favor of vertex/face positions. If your start has violated either of these, you may as well disassemble and restart from scratch.

2.10 Summary exercises

You should have learned about:

- Basics about sets, the power set of a set and its cardinality
- Set operations – how to build new sets from existing ones
- Set identities: basic laws, distributive laws, De Morgan's laws
- How to prove set identities using the double inclusion methods
- What is Russell's paradox and why it is a paradox

1. Compute $|\{n/m \in \mathbb{Q} \mid n \in \mathbb{N}; m \in \mathbb{N}; 1 \le n \le m \le 3\}|$.

2. Use the bracket notation to describe each of the following sets.
   a) Odd integers.
   b) Integers whose remainder is 1 when divided by 3.
   c) Rational numbers whose denominator can be expressed as a power of 2.

3. Consider the set $X = \mathbb{D} \cup \mathcal{P}(\mathbb{D})$. Find a subset with cardinality $2\binom{8}{3} + 13$.

4. Let $S = \{z^2 - 4z + 44 \mid z \in \mathbb{Z}\}$. Prove that $S \subseteq \mathbb{N}$.

5. Let $A, B, C,$ and D be subsets of $\mathbb{Z}$. Show that

$$(A \cap B \cap C) \subseteq (A \cup D) \cap (B \cup D) \cap (C \cup D).$$

Give an example to show that

$$(A \cup D) \cap (B \cup D) \cap (C \cup D) \subseteq (A \cap B \cap C)$$

need not be true.

6. Let $A, B,$ and C be sets. Show using the set identities that

$$(A \cup B) \cap (C \cup A^c) = (A \cap C) \cup (B \cap C) \cup (A^c \cap B).$$

7. Let $A, B,$ and C be sets. Show by the double inclusion method that

$$(A \cup B) \cap (C \cup A^c) = (A \cap C) \cup (B \cap C) \cup (A^c \cap B).$$

[Hint: For one direction it helps to consider the cases $x \in A$ versus $x \notin A$.]

8. Let $X, Y,$ and Z be sets. Prove using the double inclusion method that

$$X \cup (Z \cup Y) = (Y \cup X) \cup Z.$$

9. Let A and B be sets. Prove carefully by the double inclusion method that

$$A^c \cap B^c = (A \cup B)^c.$$

10. Let X, Y, and Z be sets. Prove using the double inclusion method that

$$X \cap (Y \cup Z) = (X \cap Y) \cup (X \cap Z).$$

11. Let X, Y, and Z be sets. Prove using the double inclusion method that

$$X \cup (Y \cap Z) = (X \cup Y) \cap (X \cup Z).$$

12. Let A and B be sets. Either prove carefully by the double inclusion method that

$$B \cup A^c = (A^c \cap B)^c$$

is true, or give an example of two sets for which the equation is false.

Chapter 3

Working with finite sets

3.1 Cardinality of finite sets

Each set operation which was defined in the previous chapter was associated with an observation, valid if the sets were finite, about the cardinalities of the sets involved. Here they all are gathered together:

$$|\emptyset| = 0 \tag{3.1}$$

$$\max(|A|, |B|) \leq |A \cup B| \leq |A| + |B| \tag{3.2}$$

$$0 \leq |A \cap B| \leq \min(|A|, |B|) \tag{3.3}$$

$$|A^c| = |\mathcal{U}| - |A| \tag{3.4}$$

$$|A \times B| = |A| \cdot |B| \tag{3.5}$$

$$|\mathcal{P}(A)| = 2^{|A|} \tag{3.6}$$

$$|\mathcal{P}_k(A)| = \binom{|A|}{k} \tag{3.7}$$

$$2^n = \sum_{k=0}^{n} \binom{n}{k} = \sum_{k=0}^{n} |\mathcal{P}_k(\{1, 2, \ldots, n\})| \tag{3.8}$$

$$0 = \sum_{k=0}^{n} (-1)^k \binom{n}{k} = \sum_{k=0}^{n} (-1)^k |\mathcal{P}_k(\{1, 2, \ldots, n\})|, \quad n > 0 \tag{3.9}$$

These results were justified by some application of the multiplicative principle, and at this point you should not only find each of them familiar, but be able to justify each of them to yourself with an explanation that is convincing to you. For instance, some people like to think of the last two as applications of the binomial theorem. Eq. (3.8) can also be understood as saying that the number of subsets of an n-set can be determined by adding up the number of subsets of the different possible cardinalities.

Eq. (3.9) states that, for any non-empty set, the number of subsets of even cardinality is always precisely balanced by the number of subsets of odd cardinality. This fact can be justified by a golden argument. If the set is non-empty, there is one element which can be distinguished as the golden one. This golden element allows us to pair each subset that contains the golden element to the

Discrete Mathematics With Logic. https://doi.org/10.1016/B978-0-44-318782-7.00008-3

subset obtained by removing that golden element. In this match, one subset has even cardinality, and the other odd, so the two types of subsets are in balance.

Exercises

1. Let X and Y be sets. Suppose $2^5 \leq |X| \leq 2^6$ and $2^4 \leq |Y| \leq 2^7$.
   a. What can you conclude about $|X \cup Y|$?
   b. What can you conclude about $|X \cap Y|$?
2. Let $A = \{1, 2, 3, 4\}$, $B = \{a, b, c, d, e\}$.
   a. Compute $|\mathcal{P}(\mathcal{P}_2(A) \times (B \times A))|$.
   b. Compute $|\mathcal{P}_2(\mathcal{P}(A) \times (B \times A))| + |\mathcal{P}_3(A \cap B)|$.
   c. Compute $\sum_{k=0}^{|A \cup B|} (-1)^k |\mathcal{P}_k(A \cup B)|$.
3. Find the 10th and 11th rows of Pascal's triangle. Verify that the sum of all the entries in each row is a power of 2, and that the alternating sum of each row is 0.

3.2 Bit vectors and ordering subsets

The formula $|\mathcal{P}(A)| = 2^{|A|}$ was justified using the multiplicative principle by associating each subset of the sequence of answers, for each element of A, to the query of membership. We can go further with this idea and take this sequence of answers as an encoding of that subset.

So, for example, if we have four flavors, chocolate, vanilla, strawberry, and cherry, then there are 2^4 subsets of those four flavors. If we query each subset about the membership of each of the four flavors *in the order listed* above, then subset {vanilla, strawberry} would be recorded by the answer sequence no-yes-yes-no. The awkwardness is eased by recording "yes" and "no" with the bits 1 and 0, respectively, so {vanilla, strawberry} could be recorded as $(0, 1, 1, 0)$, or even more compactly as 0110. If instead we query the subsets about the flavors alphabetically, then the same subset would be recorded as 0011.

A *bit vector* of length n is a sequence of n bits, and may be notated in sequence notation, like the 9 bit sequence $(0, 1, 1, 0, 1, 1, 1, 1, 0)$, or with the parentheses and commas suppressed for clarity, like 011011110.

Given a finite set A, with $|A| = n$ and a particular ordering on the elements of A,

$$A = \{a_1, a_2, a_3, \ldots, a_n\}, \quad a_i < a_j \text{ if } i < j,$$

the *bit vector of a subset* $X \subseteq A$ is $(B(a_n), B(a_{n-1}), \ldots, B(a_1))$ with $B(a_i) = 1$ if $a_i \in X$, and $B(a_i) = 0$ otherwise.

As anticipated, the bit vector depends on the ordering chosen. It is also to be noticed that the bit-vector associated with a subset seems to go backwards. You'll see why in the next example.

The bit vectors of length n are naturally associated with the binary numbers between 0 and $2^n - 1$. This association gives us, along with our subset encoding, a natural and useful ordering on those subsets.

Example 3.1. Let $A = \{a, b, c\}$. There are $2^3 = 8$ subsets in $\mathcal{P}(\{a, b, c\})$ and so the 2^3 bit vectors of length 3 are associated and ordered as follows:

$$\varnothing: \quad 000$$
$$\{a\}: \quad 001$$
$$\{b\}: \quad 010$$
$$\{a, b\}: \quad 011$$
$$\{c\}: \quad 100$$
$$\{a, c\}: \quad 101$$
$$\{b, c\}: \quad 110$$
$$\{a, b, c\}: \quad 111$$

As you can see, the ordering on the bit vectors, thought of as binary numbers, naturally starts with 0 recording $\varnothing$, which we will regard as the "least", and proceeds to 001, recording a subset with the single element a, the least in the ordering on A. If we had ordered the queries $(B(a), B(b), B(c))$, then the set following the empty set would be $\{c\}$, and that would seem backwards to most people, though not in any sense incorrect. $\quad \diamond$

Example 3.2. What is 1776th subset of $\mathbb{A}$ in the bit vector ordering, using the alphabetical ordering of $\mathbb{A}$?

For this, we naturally want to first either recall or compute that 1776 in binary is 11011110000. Written as a binary number, 1776 can be written with only 11 bits, but as a bit vector for subsets of $\mathbb{A}$ one should write 1776 as a string of length $26 = |\mathbb{A}|$, 00000000000000011011110000, and that bit vector encodes the subset $\{e, f, g, h, j, k\}$. The next three subsets following that one are $\{a, e, f, g, h, j, k\}$, $\{b, e, f, g, h, j, k\}$, and $\{a, b, e, f, g, h, j, k\}$. Notice that the answer depended on our decision, made earlier, to always order from 0. So the first subset is $\{a\}$, not $\varnothing$. The zeroth subset is $\varnothing$. (If you try to rewrite our description of the encoding to accommodate ordering from 1, then you will appreciate the simplification.) $\quad \diamond$

When sets were defined, it was emphasized that membership in a set imposed no ordering on the elements, and you might have expected that order would not play a role in set theory. As you have now seen, it plays in fact a key role. The fundamental assumption that the elements come with no ordering *a priori* leaves one free to consider not just one hardwired order, but any convenient or useful order; or perhaps several at once, comparing and contrasting them.

Exercises

1. List the sixteen subsets of $\{a, b, c, d\}$ in bit-vector order, using the usual ordering on the letters.

2. List the elements of $\mathcal{P}_2(\{1, 2, 3, 4, 5\})$ in bit vector order. Give the subsets, the bit vectors, and the decimal equivalents.
3. What is the 999th subset of $\{0, 1, 2, 3, 4, 5, 6, 7, 8, 9\}$ in bit vector order, (starting from 0 for $\emptyset$) using the usual ordering on the set of digits? Give the next 6 subsets as well.

3.3   Inclusion/exclusion

The cardinality of a union of two finite sets A and B, follows

$$\max(|A|, |B|) \leq |A \cup B| \leq |A| + |B|$$

with equality on the left if one set is a subset of the other, and equality on the right if the two sets A and B are *disjoint*, that is, that $A \cap B = \emptyset$. In between those two extremes, the two sets just have a non-empty intersection, and every element in that intersection is double counted in the expression $|A| + |B|$. That observation of double counting leads us to the identity

$$|A \cup B| = |A| + |B| - |A \cap B|.$$

Here is the formula for the cardinality of the union of three sets,

$$|A \cup B \cup C| = |A| + |B| + |C| - |A \cap B| - |A \cap C| - |B \cap C| + |A \cap B \cap C|,$$

which can be justified by using the top formula twice, along with the associative law of set union. (Try it if you are interested.) Alternatively, we can consider which elements are over-counted in $|A| + |B| + |C|$. The expression $-|A \cap B| - |A \cap C| - |B \cap C|$ removes all the over-counts, no matter what the cause, with the only problem being that those elements in all three sets have been removed three times. Those same elements were included three times in $|A| + |B| + |C|$, so excluding them three times leaves them totally uncounted, that is until the last term $|A \cap B \cap C|$ includes them once in the end. So the 7-term expression counts each element exactly once overall. This justification gives the formula its name, the *inclusion/exclusion principle*, called a principle and not a formula by virtue of its use in solving practical and theoretical problems.

There is an *inclusion/exclusion principle* for any number of sets. In general, for n finite sets $A_1, A_2, \ldots A_n$, we have

$$|A_1 \cup A_2 \cup \cdots \cup A_n| = |A_1| + |A_2| + \cdots |A_n|$$
$$- [|A_1 \cap A_2| - |A_1 \cap A_3| - |A_2 \cap A_3| - \cdots |A_{n-1} \cap A_n|]$$
$$+ [|A_1 \cap A_2 \cap A_3| - |A_1 \cap A_2 \cap A_4| - \cdots |A_{n-2} \cap A_{n-1} \cap A_n|]$$
$$\vdots$$
$$+ (-1)^{n+1}|A_1 \cap A_2 \cap \cdots \cap A_n|$$

and if you don't see how to fill in the "...", here is the principle of inclusion/exclusion written more precisely:

$$\left| \bigcup_{k=1}^{n} A_k \right| = \sum_{\emptyset \neq I \subseteq \{1,2,3,...,n\}} (-1)^{|I|+1} \left| \bigcap_{i \in I} A_i \right| \tag{3.10}$$

In this form, it is a bit intimidating. Let us leave it and a general explanation for a bit. First let's see how it is used. Student exercises mostly use the principle with at most four sets.

Even then, you may think that the inclusion/exclusion formula is just ridiculous. A simple count should tell you that, in computing $|A_1 \cup A_2 \cup \cdots \cup A_n|$ by inclusion/exclusion you would have to accurately compute the cardinalities of $2^n - 1$ different sets, and do all the accounting assembling the results, and not making any sign error. Even for $n = 3$ that is seven intersections, the cardinalities of seven sets to compute, seven terms to keep track of. How could that ever be simpler than simply computing the cardinality of $|A_1 \cup A_2 \cup \cdots \cup A_n|$ directly?

Example 3.3. How many six letter words on $\mathbb{A}$ either start with an a, like *abcabc*, or end with a b, like *xxbbbb*, or repeat in pairs, like *xyxyxy*?

As individual problems, each is a simple exercise in the multiplicative principle. But by accepting words which satisfy any one of the conditions, we cannot just solve it as three separate problems because the conditions interact. But we can apply inclusion/exclusion, the first step being not to start computing, but to define the sets which transform the problem into one of set cardinality. Let A be the set of six letter words on $\mathbb{A}$ which start with an a, let B be the set of six letter words on $\mathbb{A}$ which end with letter b, and let C be the set of six letter words on $\mathbb{A}$ which repeat in pairs. We can now compute $|A|$, $|B|$ and $|C|$ by the multiplicative principle: $|A| = 26^5$, $|B| = 26^5$, and $C = 26^2$. Better than that, the intersections we need can also be so attacked: $|A \cap B| = 26^4$, $|A \cap C| = 26$, $|B \cap C| = 26$. Even $A \cap B \cap C$ has a computable cardinality, $|A \cap B \cap C| = 1$. Now inclusion/exclusion gives the final result:

$$26^5 + 26^5 + 26^2 - 26^4 - 26 - 26 + 1.$$

So, actually, we were quite happy to trade the single complicated union cardinality problem for the 7 routine intersection cardinality problems. ◇

This doesn't exactly refute the objection above. That objection noticed that applying inclusion/exclusion would involve us in a *combinatorial explosion*, that is, a situation where each incremental increase in the size of the problem multiplies our work in solving it by an unyielding factor. That explosion is there, and using inclusion/exclusion does not cure that explosion at all, but at least it does help us to manage it as efficiently as can be expected.

Now, to justify the general result in Eq. (3.10) we fix an element x and, to examine how often x is included and excluded, we consider the set $B = \{X \in \{A_1, A_2, \ldots, A_n\} \mid x \in X\}$ of those A_i which contain x. The expression in Eq. (3.10) counts x only in intersections containing only elements of B. The expression counts x once with a positive sign for every non-empty subset of B with an odd cardinality, and once with a negative for each non-empty subset of B with an even cardinality. Adding up all those inclusions and exclusions doesn't give 0 as you might think from the alternating row sum of Pascal's triangle, since the empty set, an even subset, is excluded. So x is counted overall exactly once.

Also, using De Morgan's laws, you can discover an analogous formula which rewrites the cardinality of an intersection of many sets as an alternating sum of cardinalities of unions. The formula, which you will discover, will be true but not very useful in solving practical problems, at least not useful enough to give it a catchy name.

Exercises

1. How many strings of length 5 on $\{a, b, c, d, e\}$ either start with a, end in a, or have all the same letter?
2. How many strings of length 5 on $\{a, b, c, d, e\}$ either start with three identical characters, end with three identical characters, or have c exactly in the middle?
3. How many strings of length 25 on $\{a, b, c, d, e, \ldots, z\}$ either start with 5 identical characters, repeat every 5, (like $acadcacadcacadcacadcacadc$) or are palindromic, that is, read the same forwards and backwards (like $aabbbbbbbccacacaccbbbbbbbaa$)?

3.4   Multiple Cartesian products and strings

For the collection and organization of information, the binary Cartesian product of sets is theoretically sufficient. For many applications, however, the binary form is inconvenient since the levels of nested parentheses may serve no purpose, but still require careful attention. The multiple product is a valuable alternative.

Definition 3.4. Let $A_1, \ldots A_N$ be sets. The N-fold Cartesian product of those sets is defined as

$$A_1 \times A_2 \times \cdots \times A_N = \{(a_1, a_2, a_3, \ldots, a_n) \mid a_i \in A_i \text{ for all } 1 \le i \le N\}. \quad \spadesuit$$

So the elements of the product are N-tuples, or sequences of N elements where the coordinates, or entries, are matched one by one to the set of which they are a member. By the multiplicative principle, if all the sets are finite, then

$$|A_1 \times A_2 \times \cdots \times A_N| = |A_1| \cdot |A_2| \cdots |A_N|.$$

Example 3.5. Consider the product $\{\mathbb{Z} \times \{+, -, \times, \div\} \times \mathbb{Z} \times \{=, \leq, <, \geq, >\} \times \mathbb{Z}\}$. Elements of this set, like $(1, +, 1, =, 2)$ model simple arithmetic statements, and of course include as well such gems as $(5, \times, 5, \leq, 0)$, but not $(1, 1, +, =, 2)$. $\diamond$

It is still possible that the individual elements of a multiple Cartesian product are themselves tuples. If you wanted to model paths joining 5 locations on an 8×8 grid, you might first set $X = \{1, 2, \ldots 8\} \times \{1, 2, \ldots 8\}$ to establish the locations, and then use $X \times X \times X \times X \times X = X^5$ to model the paths, with a typical element in the set being $((1, 1), (3, 4), (8, 7), (3, 4), (8, 6))$. Note that for this structure, the entries in the tuple are both ordered (left to right) and may have multiplicities.

If there is no confusion, it is common in a multiple Cartesian product, as with the bit vectors, to suppress the commas and the parentheses to give a cleaner look. We would certainly do this with the elements in the arithmetic example, writing $5 + 5 = 10$ instead of $(5, +, 5, =, 10)$, and even in the grid path example one can imagine writing $(1, 1)(3, 4)(4, 4)(3, 3)(8, 6)$ or even 1134443386 without difficulty, but if it was a 16×16 grid, it would be difficult to know what to do with 1111111111111111.

An important special case is the N-fold Cartesian product of a set A with itself, $A \times A \times A \cdots \times A = A^N$. Please be warned that the exponent here is a bit perilous, since $(A^5)^2 \neq (A^2)^5 \neq A^{10}$ as sets, although by the multiplicative principle their cardinalities satisfy $|(A^5)^2| = |(A^2)^5| = |A|^{10}$.

For a finite set, Σ, an element of the product Σ^N is termed a *string* of length N on the *alphabet* Σ, and the set of all strings on Σ is

$$\Sigma^* = \bigcup_{k=0}^{\infty} \Sigma^k.$$

The use of the $*$ above is called the *Kleene Star*. Do not be misled by the ∞ symbol: k takes every integer value from 0 up to *but not including* ∞, so every string in Σ^* belongs to one of the sets A^k, and therefore must have length k. That is, every string has finite length. Another thing to notice, at the other end of the union, is the case when $k = 0$. Σ^0 is defined to be the *empty string*, commonly denoted by ϵ or λ or the empty tuple (). Strings, beyond their mathematical applications, are a fundamental object of study in theoretical computer science, modeling not only the typical input and output of computer programs, but the programs themselves.

For a completely different application, while it is sufficient theoretically to distribute two sets at a time, using $A \cup (B \cap C)$ or $A \cap (B \cup C)$, practically it is much better to have a more general rule describing how to handle multiple terms at once. Up until now, however, we have not had the notation to write the rule down efficiently. The multiple Cartesian product is the last piece that we need.

Theorem 3.6 (General Distributive Law for Sets). *Let* $X_1, \ldots X_n$ *be non-empty sets such that each element of each set is itself a set. Then the following hold;*

$$\bigcup_{i=1}^{n} \left[\bigcap_{X \in X_i} X \right] = \bigcap_{(X_1, \ldots, X_n) \in X_1 \times \cdots X_n} \left[\bigcup_{i=1}^{n} X_i \right], \qquad (3.11)$$

$$\bigcap_{i=1}^{n} \left[\bigcup_{X \in X_i} X \right] = \bigcup_{(X_1, \ldots, X_n) \in X_1 \times \cdots X_n} \left[\bigcap_{i=1}^{n} X_i \right]. \qquad (3.12)$$

This useful rule uses three different forms of notation for general union and intersection. The easiest are the two on the left, where the multiple union and intersections are indexed by either a set of consecutive natural numbers, $\{1, 2, \ldots, n\}$, or a more general set. On the right hand side the format of the 'dummy variable' is used to specify the required term.

Example 3.7. $X_1 = \{A, B, C\}$, $X_2 = \{D\}$, and $X_3 = \{A, E\}$, the left side of Eq. (3.11) gives

$$\bigcup_{i=1}^{3} \left[\bigcap_{X \in X_i} X \right] = \left[\bigcap_{X \in X_1} X \right] \cup \left[\bigcap_{X \in X_2} X \right] \cup \left[\bigcap_{X \in X_3} X \right]$$
$$= (A \cap B \cap C) \cup D \cup (A \cap E).$$

Decoding the right hand side gives

$$\bigcap_{(X_1, X_2, X_3) \in X_1 \times X_2 \times X_3} \left[\bigcup_{i=1}^{3} X_i \right] = \bigcap_{(X_1, X_2, X_3) \in X_1 \times X_2 \times X_3} (X_1 \cup X_2 \cup X_3)$$
$$= (A \cup D \cup A) \cap (A \cup D \cup E) \cap (B \cup D \cup A)$$
$$\cap (B \cup D \cup E) \cap (C \cup D \cup A) \cap (C \cup D \cup E),$$

exactly what is implied by algebra and the basic distributive law.   ◇

As remarked, the general distributive law follows from the basic laws, and the commutativity and associativity of the intersection and union. It also is fairly straightforward to show it directly with the double inclusion method.

Exercises

1. Let $A = \{1, 2, 3\}$.
   a. Find 5 elements in $A^5 \times A \times \mathcal{P}(A)$.

  b. What is the cardinality of $A^5 \times A \times \mathcal{P}(A)$?
2. Consider the set $((\{a, b, c\}^2)^2)^2$.
   a. Find 5 elements of the set. Do not suppress any commas or parentheses.
   b. What is the cardinality of the whole set?
3. Find 5 elements of $\{P, B, T, D\} \times \{u, o\} \times \{s, z\}^2 \times \{a, e\} \times \{y, t\}$ suppressing all commas and parentheses. What is the cardinality of the whole set?

3.5 Lexicographical order

Lexicographic order on multiple products

Given two strings of the same length on $\mathbb{A}$, we know how to order them alphabetically, by comparing the first letters. If the first two letters are different, the string starting with the larger letter is larger. If the first two letters are the same, then we compare the second letters, from left to right, comparing the second letters and so on. Alphabetic order is also called dictionary order, or *lexicographic order*. So "cat", precedes "dog", and "dog" precedes "dot". The same method extends to strings of fixed length, and multiple Cartesian products in general.

Given a collection of sets A_i, and an ordering on each set, it is possible to extend those orderings lexicographically to the set $A_1 \times A_2 \times \cdots \times A_N$. Comparing two elements in this set lexicographically, $(x_1, x_2, \ldots, x_N)$ versus $(x'_1, x'_2, \ldots, x'_N)$, we define the *lexicographic order on* the elements, writing $(x_1, x_2, \ldots, x_N) < (x'_1, x'_2, \ldots, x'_N)$, if there is a $1 \leq k \leq N$ with $x_i = x'_i$ for all $i < k$ and $x_k < x'_k$.

Under this definition, the order on the elements of a multiple product depends only on the relative order of the first differing coordinate, just as do the words in the dictionary.

So for example, set $\mathbb{D} \times \{-, +\} \times \mathbb{D} \times \{\leq, =, \geq\} \times \mathbb{D}$, which we used to model first grade arithmetic, contains 6000 elements. If we take the usual ordering on $\mathbb{D}$ and the order in the listing for the other two sets, the first element in lexicographic order is $0 - 0 \leq 0$, and the last one is $9 + 9 \geq 9$, both of which happen to be true. For any choice of the first four coordinates, there are 10 choices of the final coordinate, ordered sequentially in lexicographic order, say from $5 + 7 = 0$ to $5 + 7 = 9$, all of which are false, but that is not the issue. Incrementing $5 + 7 = 9$, we have $5 + 7 \geq 0$, and then the sequence of 10 restarts. Incrementing any element, say $5 + 7 \geq 9$, involves moving ahead in the rightmost coordinate for which that is possible, $5 + 8\_\_$, and resetting all the coordinates further to the right to the least element in the relevant set, $5 + 8 \leq 0$. We see that the lexicographic list is organized in the same way that the list of natural numbers in base 10 is organized, except that instead of the columns being sequences of 10 identical characters in the 10's place, and 10^2 identical characters in the 10^2's place, etc., there will be 10 identical characters in the $\{\leq, =, \geq\}$ place, $3 \cdot 10$ identical characters in the middle $\mathbb{D}$ place, $10 \cdot 3 \cdot 10$ identical characters in the middle $\{-, +\}$ place, etc. And it is no surprise that the backwards method of computing a number to a different base adapts easily to this situation.

Theorem 3.8. *Suppose we have a collection of N sets $A_0, \ldots A_{N-1}$ and an ordering on each set, and a number $m < |A_0| \cdots |A_{N-1}|$. Let $\{m_i\}$ and $\{r_i\} $ be defined by setting $m_0 = m$, and the divisions by $|A_i|$ with remainder $m_i = m_{i+1}|A_i| + r_i$. Then the m'th element of $A_{N-1} \times \cdots \times A_0$ in lexicographic order is $a_{N-1} \cdots a_0$ with each a_i the r_ith element of A_i, (ordering from the 0th).*

On the other hand, given an element $a_{N-1} a_{N_2} \cdots a_1 a_0 \in A_{N-1} \times \cdots \times A_0$, that element has position

$$r_{N-1}|A_{N-2}| \cdots |A_0| + \cdots + r_3|A_2||A_1||A_0| + r_2|A_1||A_0| + r_1|A_0| + r_0,$$

with each a_i the r_ith element of A_i, ordering from the 0th.

If we now want to determine the 1776th element along the list of 6000 elements of $\mathbb{D} \times \{-, +\} \times \mathbb{D} \times \{\leq, =, \geq\} \times \mathbb{D}$, we compute $1776 = 177 \cdot 10 + 6$, then $177 = 59 \cdot 3 + 0$, then $59 = 5 \cdot 10 + 9$, then $5 = 2 \cdot 2 + 1$, and then finally $2 = 0 \cdot 10 + 2$. The remainders in reverse order of the computations, $(2, 1, 9, 0, 6)$, record the element required: $2 + 9 \leq 6$. It is important to be careful in the final step if you are not yet used to ordering from 0; the 1st element of $\{-, +\}$ is $+$, since $-$ is the 0th. Reversing the process we would find that $5 + 9 \leq 6$ is the $2 \cdot 2 \cdot 10 \cdot 3 \cdot 10 + 1 \cdot 10 \cdot 3 \cdot 10 + 9 \cdot 3 \cdot 10 + 0 \cdot 10 + 6$th, which you can check is 1776, as required.

Lexicographic order on Σ^k

Lexicographic order on a multiple Cartesian product has a nice interpretation in the case where all the sets in the product are the same, for example when considering strings of length k and an alphabet Σ. To find the mth string in Σ^k, the divisions with remainder are all done with respect to the same number, $|\Sigma|$. In fact, we can observe that, in the special case of ordering strings of length k from an alphabet Σ, the problem is essentially the same as counting k-digit numbers in base $|\Sigma|$, (and one can even regard the ordered elements of Σ as the digits!)

Theorem 3.9 (Lexicographically Ordering Strings). *Let $\Sigma = \{a_0, a_1, \ldots a_{N-1}\}$ be a set of cardinality N with ordering $a_0 < a_1 < \cdots < a_{N-1}$ on its elements.*

Let $m = d_{N-1}N^{k-1} + d_{N-2}N^{k-2} + \cdots + d_1 N^1 + d_0 N^0$, with each d_j satisfying $0 \leq d_i < N$, be the representation of the number m as a k digit number in base N. Then the mth string in Σ^k, regarding $a_0 a_0 \cdot a_0$ as the 0th, is $(a_{d_{N-1}}, a_{d_{N-2}}, \ldots, a_{d_0})$, or more compactly $a_{d_{N-1}} a_{d_{N-2}} \cdots a_{d_1} a_{d_0}$.

Example

Let $\Sigma = \{a, b, c, d, e, f, g, h, i, j\}$ be an alphabet of 10 letters, and the usual ordering. The 5280th string of Σ^4 in lexicographic order is $fcia$, because 5280 is already represented in base 10 and the entries of $fcia$ are the 5th, 2nd, 8th, and 0th characters of Σ. The 5280th string of Σ^8 in lexicographic order is $aaaafcia$.

If we reduce the alphabet to $\Sigma' = \{a, b, c, d, e, f\}$ and want the 5280th string of length 8, we have to first convert 5280 to base 6. By the backwards method of Section 1.6 we just have to repeatedly divide by 6 and get $5280 = 40240_6$, or $5280 = 0 \cdot 6^7 + 0 \cdot 6^6 + 0 \cdot 6^5 + 4 \cdot 6^4 + 0 \cdot 6^3 + 2 \cdot 6^2 + 4 \cdot 6^1 + 0 \cdot 6^0$. That yields the string *aaaeacea*. The formula of Theorem 3.9 works just as well for the opposite problem, which is called *delisting*.

Given the string *abccebef* $\in \Sigma^8$, what position does it have on the list of strings in Σ^8 in lexicographic order? If the string *aaaaaaaa* is the 0th, *abccebef* is string $0 \cdot 6^7 + 1 \cdot 6^6 + 2 \cdot 6^5 + 2 \cdot 6^4 + 4 \cdot 6^3 + 1 \cdot 6^2 + 4 \cdot 6^1 + 5 \cdot 6^0$.

Varying the length

Decimal numbers, binary numbers, and ordinary words of varying length may all be ordered lexicographically by adapting our mathematical lexicographical order, yet the methods used are quite different. In comparing numbers of different digit length, the shorter number is padded on the left with zeros. So 69 and 5280 are compared just as 0069 and 5280, ignoring the fact that, according to first non-zero digits, $6 > 5$. Words not of the same length, however, are compared lexicographically by padding the shorter word on the right with blanks, with the blank regarded as less than any letter for purposes of individual comparison. For example, *blackie* and *blackbird* are compared via *blackbird* < *blackie__*, not *__blackie* < *blackbird*.

Exercises

1. Let $B = \{a, b\}$.
   How many strings are between *abbbb* and *baaaa* in the lexicographic order on all strings of length 5 on B?
2. Which number is the string $\gamma\alpha\gamma\alpha\beta\epsilon\epsilon\beta\epsilon\epsilon$ in lexicographical order on the strings of length 10 on $\{\alpha, \beta, \gamma, \delta, \epsilon\}$?
3. Let $A = \{1, 2, 3\}$ and $B = \{a, b\}$. Taking the usual ordering on the elements of both sets, what is the 88th element of $A \times B \times A \times B \times A \times B \times A$ in lexicographic order?

3.6 Ordering permutations

Subsets of multiple Cartesian products

We have seen how we can list and delist the elements of a multiple Cartesian product of finite sets. In particular, we have a close relationship between the lexicographic ordering on the strings of length k on an alphabet Σ, and the k-digit numbers written in base $|\Sigma|$. What about subsets of a multiple Cartesian product? Can we solve the listing and delisting problem when the elements are restricted? For example, for the multiple Cartesian product $\mathbb{D} \times \{-, +\} \times \mathbb{D} \times \{\leq, =, \geq\} \times \mathbb{D}$ whose elements model simple digit arithmetic problems, what if we want to restrict to those elements for which the mathematical expression is true?

$$T = \{x \in \mathbb{D} \times \{-, +\} \times \mathbb{D} \times \{\leq, =, \geq\} \times \mathbb{D} \mid x \text{ is true}\}$$

So $9 + 1 > 0 \in T$, but $2 + 2 = 5 \notin T$. So we could ask, how many elements of T are ahead of $9 + 1 > 0$ in lexicographic order? Or we could ask which digit fact is 1776th in the ordering? Both questions look very difficult, in fact, it looks difficult to determine if there even are 1776 elements in T. In general, in order to get a complete answer for a subset of a multiple Cartesian product, that subset would have to be quite regularly constructed.

Permutations

The set of permutations of the six characters in $\Sigma = \{a, b, c, d, e, f\}$ can be regarded as a subset of Σ^6, specifically the subset

$$P(\Sigma) = \{\sigma_1\sigma_2\sigma_3\sigma_4\sigma_5\sigma_6 \in \Sigma^6 \mid \sigma_i \neq \sigma_j \text{ for all } i \neq j\}.$$

If $P(\Sigma)$ is ordered lexicographically, the 0th element is $abcdef$, followed by $abcdfe$, etc., until the $(6! - 1)$th element $fedcba$.

abcdef	abdcef	abecdf	abfcde	acbdef	acdbef	acebdf	fedabc
abcdfe	abdcfe	abecfd	abfced	acbdfe	acdbfe	acebfd	fedacb
abcedf	abdecf	abedcf	abfdce	acbedf	acdebf	acedbf ...	fedbac
abcefd	abdefe	abedfc	abfdec	acbefd	acdefb	acedfb	fedbca
abcfde	abdfce	abefcd	abfecd	acbfde	acdfbe	acefbd	fedcab
abcfed	abdfec	abefdc	abfedc	acbfed	acdfeb	acefdb	fedcba

You may take a general element of $P(\Sigma)$, say $dfbeca$, and ask what follows it in $P(\Sigma)$ in lexicographic order. You cannot increment just the final a, because it is forced by the first part of the string. You cannot increment the c next to it because, fixing the characters to the left, it is already the highest value among a and c. You cannot increment the e next door for the same reason. It is the b which may be incremented, holding the initial df fixed, to c. So the next permutation after $dfbeca$ in lexicographic order is $dfcade$, taking the leftover letters $\{a, d, e\}$ in order. (After that come $dfcaed$, $dfcdae$, $dfcdea$, ...). Observing the list above we see the pattern that, reading from right to left, the rightmost coordinate has no freedom at all, its neighbor to the left has two alternatives it must cycle through, and its neighbor to the left has three alternatives it must cycle through in pairs, so *its* neighbor to the left must cycle through its four possibilities in groups of $2 \cdot 3 = 3!$, etc.

How many permutations are ahead of a general permutation $\sigma_1\sigma_2\sigma_3\sigma_4\sigma_5\sigma_6$? We ask first how many permutations were on the list before σ_1 appeared in the first position? Each element of $\{\sigma_2, \sigma_3, \sigma_4, \sigma_5, \sigma_6\}$ which is smaller than σ_1 occupied the first position ahead of σ_1, and for each one, there were $(6 - 1)!$ permutations. So if we set r_1 to be the number of elements in $\{\sigma_2, \sigma_3, \sigma_4, \sigma_5, \sigma_6\}$ less than σ_1, there were $r_1(6-1)!$ permutations before σ_1 appeared in first position. After that, how many permutations were there before σ_2 appeared in the second spot? With σ_1 fixed in the front, each element of $\{\sigma_3, \sigma_4, \sigma_5, \sigma_6\}$ which is less than σ_2 had to occupy second position first, and for each one there were $(6-2)!$ permutations. So, setting, r_2 to be the number of element of $\{\sigma_3, \sigma_4, \sigma_5, \sigma_6\}$ which are less than σ_2, we have $r_2(6-2)!$ additional permutation ahead of $\sigma_1\sigma_2\sigma_3\sigma_4\sigma_5\sigma_6$. The argument continues in this manner, computing r_3, r_4, r_5. The final result is $r_1(6-1)! + r_2(6-2)! + r_3(6-1)! + r_4(6-4)! + r_5(6-5)!$.

Here is the general result, however, it will look nicer, and help us later, if the permutation is indexed in the other direction, starting from 0.

Theorem 3.10. *Let Σ be a finite ordered set of N elements, and let $\sigma_{N-1}\cdots\sigma_1\sigma_0$ $\in P(\Sigma) \subseteq \Sigma^N$ be a permutation of Σ. Set r_i to be the number of elements in $\{\sigma_0, \ldots, \sigma_{i-1}\}$ which are less than σ_i in the ordering on Σ. Then the number of permutations in $P(\Sigma)$ which are less than $\sigma_{N-1}\cdots\sigma_0 \in P(\Sigma) \subseteq \Sigma^N$ in lexicographic order is*

$$r_{N-1} \cdot (N-1)! + r_{N-2} \cdot (N-2)! + \cdots + r_1 \cdot 1! + r_0 \cdot 0!.$$

Notice that if we compute the number of elements ahead of $\sigma_{N-1}\cdots\sigma_1\sigma_0 \in P(\Sigma) \subseteq \Sigma^N$ to be k, then $\sigma_{N-1}\cdots\sigma_1\sigma_0 \in P(\Sigma) \subseteq \Sigma^N$ is the kth permutation on a lexicographically ordered list of the elements of $P(\Sigma)$ since natural order is our 0th permutation (just as the day you are born is your 0th birthday and on your 20th birthday you have previously lived 20 years).

Example 3.11. How many permutations of $\{0, 1, 2, 3, 4, 5, 6, 7, 8, 9\}$ are ahead of 5824971063?

☩

There are 5 numbers ahead of 5, $r_9 = 5$; and there are 7 numbers ahead of 8 excluding 5, $r_8 = 7$; and there are 2 numbers ahead of 2 excluding 5 and 8, $r_7 = 2$; and continuing $r_6 = 3$, $r_5 = 5$, $r_4 = 3$, $r_3 = 1$, $r_2 = 0$, $r_1 = 1$, and of course $r_0 = 0$. Altogether, the number of permutations is

$$5 \cdot 9! + 7 \cdot 8! + 2 \cdot 7! + 3 \cdot 6! + 5 \cdot 5! + 3 \cdot 4! + 1 \cdot 3! + 0 \cdot 2! + 1 \cdot 1! + 0 \cdot 0!$$

◇

Exercises

1. List all 24 permutations of $\{a, b, c, d\}$ in lexicographical order.
2. How many numbers with 10 distinct decimal digits are less than 8,214, 596,073? (The smallest such number is 0,123,456,789.)
3. Starting with 8,214,596,073, what are the next 12 numbers with distinct digits in order?

3.7 Delisting permutations[†]

We have seen that, if a finite set Σ, $|\Sigma| = N$ has an ordering, then the $N!$ permutations of that set may be regarded as a subset of $\Sigma^N = \Sigma^{|\Sigma|}$ and sorted in lexicographic order. We also saw in Theorem 3.9 how, given a permutation, one can determine quickly its place on the lexicographic list without running through the whole list.

In this section we want to show how to reverse the process, to see how to find for any number $0 \le m < N!$ the mth permutation on the lexicographic list. This is done in a two stage process. First, we discover how, given m, to find the numbers $r_0, r_1, \ldots$ of Theorem 3.9, and then, given those numbers, how to assemble the permutation.

We will use the same type of backwards procedure we used for multiple Cartesian products. Remember, lexicographic order is biased to the left of the string, and our natural bias is to consider the terms in the string from left to right, but it is preferable to index from right to left, $\sigma_{N-1} \cdots \sigma_1 \sigma_0$ just as is done with decimal numbers. For the final term in the permutation, since it is forced by the ones before, r_0 is always 0. The work starts with r_1. Given what is to its left, the 1st term from the right has two alternatives, with the permutations in even positions on the lexicographical list taking the smaller alternative, and those in odd position taking the larger. So if we use division with remainder, setting $m = m_1$, the division by 2 yields r_1 via $m_1 = 2 \cdot m_2 + r_1$. Moving on to the left, the next term has three alternatives, these appear in pairs going down on the list, smallest to largest, so again division with remainder $m_2 = 3 \cdot m_3 + r_2$ gives us m_3 and r_2. We continue in this fashion generating $r_3, \ldots r_{N-1}$.

To make the general description smoother, we can set $m_0 = m = m_1$ make an initial division by 1, with remainder $r_0 = 0$. So we are following exactly the procedure for multiple Cartesian products in the situations where the cardinalities of the sets decrease steadily from N to 1. As a side benefit, we get the interesting formula for m in terms of the remainders r_i:

$$m = r_{N-1} \cdot (N-1)! + r_{N-2} \cdot (N-2)! + \cdots + r_2 \cdot 2! + r_1 \cdot 1! + r_0 \cdot 0!$$

Here is the general description.

Theorem 3.12. *For the mth permutation of Σ in lexicographic order, in order to find the numbers r_i of Theorem 3.9, set $m_0 = m$, and perform $|\Sigma|$ divisions with remainder, $m_i = (i+1) \cdot m_{i+1} + r_i$.*

Example 3.13. To find the 1776th permutation of $\{a, b, c, d, e, f, g, h, i, j\}$, we do

$$
\begin{array}{ll}
1776 = 1 \star 1776 + 0 & \quad 14 = 6 \star 2 + 2 \\
1776 = 2 \star 888 + 0 & \quad 2 = 7 \star 0 + 2 \\
888 = 3 \star 296 + 0 & \quad 0 = 8 \star 0 + 0 \\
296 = 4 \star 74 + 0 & \quad 0 = 9 \star 0 + 0 \\
74 = 5 \star 14 + 4 & \quad 0 = 10 \star 0 + 0
\end{array}
$$

in ten divisions. This shows that the only remainders which are non-zero are $r_4 = 4$ and $r_5 = r_6 = 2$, and we may check that, indeed, $1776 = 2 \cdot 6! + 2 \cdot 5! + 4 \cdot 4!$.

To complete the process, we have to find, given $r_0, r_1, \ldots$, the desired permutation $\sigma_{N-1} \sigma_{N-2} \cdots \sigma_1 \sigma_0$. So far we have been working backwards, from right to left, but this will no longer help us. The fact that $r_0 = 0$ only tells us that σ_0 is forced by the coordinates to the left. We need to find those first, and so we better work now from left to right.

For the $N - 1$th position, take the r_{N-1}th element from the N elements in Σ. For the $N - 2$th position, take the r_{N-2}th element from the $N - 1$ elements remaining, and at each stage both fill the coordinate position, and remove that value from the next and all subsequent choices.　　◇

Example 3.14. Continuing with our example, we have $r_9 = r_8 = r_7 = 0$, so we start the permutation with abc, very much at the front of our dictionary, which is to be expected since it has $10! = 3,628,800$ entries, and we are only looking for the 1776th. The entries remaining are $\{d, e, f, g, h, i, j\}$, and $r_6 = 2$, and the 2nd from the left, ordering from the 0th, is f, so $abcf$. Next we want the 2nd from $\{d, e, g, h, i, j\}$, which is g, and form $abcfg$. Next we want the 4th from $\{d, e, h, i, j\}$ which is j, and so form $abcfgj$. All the remaining r's are 0, we put in the remaining $\{d, e, h, i\}$ in order, and form at last $abcfgjdehi$.　　◇

The general description we are leaving as an exercise. A hint is, in the previous general description, it was done by introducing the sequences of numbers r_i and m_i, here one of the sequences of objects to be introduced are strings, and the other are sets. To get you started, here is another example whose answer can be checked from the list begun in Section 3.6.

Example 3.15. What is the 32nd permutation of $\{a, b, c, d, e, f\}$? We begin by finding the sequence of remainders:

```
32 = 1 * 32 + 0        5 = 4 *   1 + 1
32 = 2 * 16 + 0        1 = 5 *   0 + 1
15 = 3 *   5 + 1       0 = 6 *   0 + 0
```

and check that $32 = 1 \cdot 4! + 1 \cdot 3! + 1 \cdot 2!$. Since $r_5 = 0$, the permutation starts a. For the next three letters, since $r_2 = r_3 = r_4$, we skip b, and form $acde$, with the last two letters $\{b, f\}$ appearing in order, $acdebf$. If you look on the list, you will find it in the correct spot, remembering that $abcdef$ is the 0th.　　◇

Exercises

1. Compute with the backwards method the 0th, 10th, and 20th permutations of $\{\alpha, \beta, \gamma, \delta\}$ in lexicographical order.
   Verify your results by listing them.
2. What is the 720th permutation of $\{a, b, c, d, e, f, g\}$ in lexicographical order, counting from 0?

3. What is the 666th natural number, counting from 0, in order of increasing size, that has 10 distinct decimal digits?
(The smallest is 0,123,456,789.)

3.8 Case study: Wolf-Goat-Cabbage

In Chapter 1 we considered the classic Wolf-Goat-Cabbage problem.

Can a man cross a river with three items, a wolf, a goat, and a basket of cabbages, in a boat which is just large enough for him to take at most one item of cargo at a time, keeping in mind that, if left unguarded, the wolf will kill the goat, and the goat will eat the cabbage.

It was introduced as an example of a discrete problem which could be studied mathematically if one cared to analyze it carefully, instead of just playing idly with it. Our example was to use the multiplicative principle to count the number of relevant puzzle positions, the number of *states* by considering the yes/no questions which asked whether each of the man, wolf, goat and cabbage was on the near side of the river or not, and concluded that the number of relevant states was at most $2^4 = 16$.

Later, when sets were introduced, we could have noticed that these yes/no questions were essentially bit vectors, and that the states could be naturally identified with one of the 2^4 subsets of {*Man, Wolf, Goat, Cabbage*}, in other words, the states could be regarded as elements of $\mathcal{P}(\{Man, Wolf, Goat, Cabbage\})$. In this encoding, the initial puzzle position is {*Man, Wolf, Goat, Cabbage*} and the happy ending is $\emptyset$.

We can also use strings to study solutions and even possible solutions and partial solutions. The alphabet will be the set $\Sigma = \{\emptyset, w, g, c\}$, which we will regard as a single action of the Viking crossing the fjord with the indicated article w, g or c, or all alone for $\emptyset$. Then any sequence of actions, any attempt of the Viking to solve the puzzle, can be regarded as an element of Σ^*. So a string starting $g\emptyset c\ldots$, would be interpreted to mean that the Viking first crosses with the goat, then returns with the goat left alone on the far side, and then crosses with the cabbages to the other side. A string starting $w\emptyset c\emptyset\ldots$ would be interpreted as the Viking crossing the fjord with the wolf and returning for the cabbage, but this string cannot be the initial segment of any solution because, according the rule, the cabbage would no longer be there.

There are other issues one encounters in applying these action strings to the problem. A string starting $g\emptyset g\ldots$ is impossible to interpret even though no mayhem has been committed because, when the man returns to the near shore, there is no goat to cross with.

You might object that Σ^* is unnecessarily huge, and that Σ^{16} is certainly sufficient by the bound on the number of states. It depends on what you want to study. That sounds persuasive if you want to find the shortest possible solution. But other people might want to allow for the puzzling Viking to start his solution $gggggggg\cdots$

We will return to the case study for more analysis, but most readers I think will have a solution in Σ^7 before then.

3.9   Case study: The Gray code

We found in the bit vectors $\mathbb{B}^N$ a very convenient encoding of the subsets of a set of cardinality N, and the encoding of those bit vectors as binary numbers gave us a familiar ordering on those subsets. But that ordering is not necessarily the most convenient. Suppose you have 15 billiard balls, and you just wanted to photograph each subset of balls, putting each subset in turn in a box on which your camera is focused. You first photograph the empty box, click, then the 1 ball alone in the box, click, swap the 1 and 2 balls, click, put the 1 ball back, click, empty the box and put in the 3 ball, click, etc. Of course, there are 2^{15} subsets, so it will take you a long time, but before every power of two, you must empty the box of all the balls you have been using so far to include just the new ball all alone. Perhaps there is a better way.

Here is a similar problem. You have a line of 15 switches on the wall, and want to pass through all possible combinations of those switches being on and off. If you use the binary number order on the bit vectors of those switches, you will quite often have to flip many switches at the same time. Can you cut that down? Can you arrange the bit vectors so that, in fact, you never have to flip more than one switch at a time?

The answer is yes. There are several ways and all the methods are based on the following trick. If you have a method for k switches, you can extend it to $k + 1$ switches by using your method on the first k switches, then flipping just the $k + 1$ switch, and then reversing your method with the first k switches, leaving the $k + 1$st switch on. Since half the positions have the $k + 1$st switch on, and half have it off, all possible positions with $k + 1$ switches are covered.

Let's try this, starting with four switches, moving from bottom to top, with initial position 1011. The second version is identical to the first, except that the switch to be flipped is indicated with an x.

```
1 1 1 1 1 1 1 1 0 0 0 0 0 0 0 0    1             x             0
0 0 0 0 1 1 1 1 1 1 1 1 0 0 0 0    0       x             x     0
1 1 0 0 0 0 1 1 1 1 0 0 0 0 1 1    1  x        x        x     x 1
1 0 0 1 1 0 0 1 1 0 0 1 1 0 0 1    1x   x    x    x    x    x   x1
```

If now we noticed that there was a fifth switch above these, say, 01011, we could extend by flipping that now, and reverse the flips of the other four:

```
0 0 0 0 0 0 0 0 0 0 0 0 0 0 0 0x1 1 1 1 1 1 1 1 1 1 1 1 1 1 1 1
1 1 1 1 1 1 1 1x0 0 0 0 0 0 0 0 0 0 0 0 0 0 0 0x1 1 1 1 1 1 1 1
0 0 0 0x1 1 1 1 1 1 1 1x0 0 0 0 0 0 0 0x1 1 1 1 1 1 1 1x0 0 0 0
1 1x0 0 0 0x1 1 1x0 0 0x1 1 1x0 0 0x1 1 1x0 0 0x1 1 1x0 0 0x1 1
1x0 0x1 1x0 0x1 1x0 0x1 1x0 0x1 1x0 0x1 1x0 0x1 1x0 0x1 1x0 0x1
```

If we follow this procedure starting with 00000, the resulting sequence of bit vectors is called the *Gray code*, and for five bits it is

```
0 0 0 0 0 0 0 0 0 0 0 0 0 0 0 0 0x1 1 1 1 1 1 1 1 1 1 1 1 1 1 1
0 0 0 0 0 0 0 0x1 1 1 1 1 1 1 1 1 1 1 1 1 1 1x0 0 0 0 0 0 0 0
0 0 0 0 0x1 1 1 1 1 1 1 1x0 0 0 0 0 0 0 0 0x1 1 1 1 1 1 1 1x0 0 0 0
0 0x1 1 1 1x0 0 0 0x1 1 1 1x0 0 0 0x1 1 1 1x0 0 0 0x1 1 1 1x0 0
0x1 1x0 0x1 1x0 0x1 1x0 0x1 1x0 0x1 1x0 0x1 1x0 0x1 1x0 0x1 1x0
0x1 1x0 0x1 1x0 0x1 1x0 0x1 1x0 0x1 1x0 0x1 1x0 0x1 1x0 0x1 1x0
```

We would like to figure out what the kth configuration of switches in the Gray code is. Let the number k be represented by bits in binary as $b_{N-1}b_{N-2}\ldots b_1 b_0$, and we would like to determine the Gray encoding $g_{N-1} g_{N-2}\ldots g_1 g_0$.

A backwards method applies here, and we consider determine first the 0th Gray bit. Notice the 0th bit repeats in pairs in binary, but repeats in groups of four in the Gray code. They don't match exactly. But the last two bits in binary repeat every four, in the order 00, 01, 10, 11, which is sufficient to determine the 0th bit of the Gray code, 00 and 11 giving 0, and 10 and 01 giving 1. A more compact description is

$$g_0 = \begin{cases} 0 & \text{if } b_0 = b_1 \\ 1 & \text{if } b_0 \neq b_1 \end{cases}$$

But the same consideration holds between g_1 and the next pair of binary bits, b_1 and b_2. The pattern takes twice as long to repeat for both, but only because it develops in identical pairs, so it is still true that

$$g_1 = \begin{cases} 0 & \text{if } b_1 = b_2 \\ 1 & \text{if } b_1 \neq b_2 \end{cases}$$

with the general encoding given by

$$g_k = \begin{cases} 0 & \text{if } b_k = b_{k+1} \\ 1 & \text{if } b_k \neq b_{k+1} \end{cases}$$

So, for example, if you want the 13th element of the Gray code, you first express 13 in binary $13 = 1101_2$, then, moving from right to left, record all the places where 0 switches to 1 and vice versa in 01101, so 01011.

Decoding from a Gray string back to a binary string will have to wait.

3.10   Case study: The forgetful waitress problem

The problem concerns a waitress in a restaurant, and a dozen customers at one of her tables. A good waitress, one who wants a big tip, will get to know her customers and try to, when she returns with the customers' meals, distribute the orders to the proper customer without asking who ordered what. Our forgetful waitress is very busy and has other things on her mind. Everybody ordered something different and she has forgotten completely who ordered what, but still she doesn't want to ask, so she distributes the orders randomly, hoping that she will guess correctly. Her probability of success is an easy exercise in the multiplicative principle. The Forgetful Waitress Problem, however, asks the probability that *nobody* gets the correct order.

✠

The number of ways to distribute the orders is 12!, the number of permutations, so we need only compute the number permutations of, say, *abcdefghiklm* such that no letter is in its correct alphabetical position.

A permutation of this type is called a *derangement*. So we are asking what fraction of permutations of 12 elements are derangements.

That problem also looks like an easy problem for the multiplicative principle, but a naive left to right approach won't work. There are 11 choices for the first position, true. But then the number of choices for second position depends. If you picked *b* for first position, a legal choice, then any of the remaining eleven letters can be placed in second position. If, on the other hand, you picked *c*, then there would be only 10 choices for second position, since both *b* and *c* would be forbidden. You can split into two cases, but then it gets even worse further on.

A clever trick is to, first of all, count the complement – the number of permutations for which at least someone gets the correct order, and secondly, to count that with inclusion/exclusion. This is a bit more difficult than our exercise problems, but follows the same pattern. First we have to define our sets. Let P_i be the number of permutations of *abcdefghiklm* such that the ith letter is fixed. So P_0 if the set of those permutations which start with a and P_{11} is the set of those which end in m. We want to count $|P_0 \cup P_1 \cup \cdots P_{10} \cup P_{11}|$.

With 12 sets, inclusion/exclusion will involve us with $2^{12} - 1 = 4,095$ cardinality of intersection problems, but don't panic! First $|P_i| = (12 - 1)!$, since one letter is fixed and the rest are permuted. There are 12 such sets, so the first positive terms in the inclusion/exclusion formula give us, altogether, $12 \cdot (12 - 1)!$.

For the pairwise intersection $|P_i \cap P_j| = (12 - 2)!$ since 2 elements are fixed, and the rest are permuted. There are $\binom{12}{2}$ terms like this, and they are all subtracted so $-\binom{12}{2}(12 - 2)!$ altogether are excluded.

Skipping ahead to the general case for an intersection of k, we have $(12 - k)!$ permutations, and there are $\binom{12}{k}$ such terms, and they are subtracted if k is even, and added if k is odd, i.e. $(-1)^{k-1}\binom{12}{k}(12 - k)!$. You were right not to panic, by gathering all the like terms together, we only have 11 terms, and not 4,095. For $k = 6$, since $\binom{12}{6} = 924$ we were able to handle almost one thousand of those 4095 sets all at once!

Let's clean up the first one, which was $12 \cdot (12 - 1)!$ and rewrite it as $(-1)^0\binom{12}{1}(12 - 1)!$, so that it fits the pattern. Now we add:

$$\sum_{k=1}^{12} (-1)^{k-1} \binom{12}{k} (12-k)!$$

Before you program that in, recall that these are the permutations we do *not* want, the ones where somebody does get the correct order. We have to subtract these from 12!, which we will write as $\binom{12}{0}(12-0)!$ to make it fit the pattern. With that, the number of derangements is

$$\sum_{k=0}^{12} (-1)^{k} \binom{12}{k} (12-k)!$$

(Note that the sum now goes from 0.)

Now, for the probability that nobody gets the ordered meal, we divide by 12! and we get

$$\sum_{k=0}^{12} (-1)^{k} \binom{12}{k} \frac{(12-k)!}{12!} = \sum_{k=0}^{12} (-1)^{k} \left[\frac{12!}{k!(12-k)!} \right] \frac{(12-k)!}{12!} = \sum_{k=0}^{12} (-1)^{k} \frac{1}{k!}$$

Now you can program it in.

But if you have taken Calculus, you might not want to, since there it is shown that

$$e^{x} = \sum_{k=0}^{\infty} x^{k} \frac{1}{k!},$$

so the number we will get for the finite sum is very close to $e^{-1} = \frac{1}{e}$, a little more than 1/3 of the time. Moreover, the more people at the table, the closer the probability that nobody receives what they ordered gets to $1/e$. Except that it doesn't, unless you have a science fiction restaurant where billions of people can order billions of different things.

But mathematically, we can still say that the fraction of permutations which are derangements approaches $1/e$ as the number of objects permuted approaches infinity.

3.11 Summary exercises

You should have learned about:

- Basic facts about cardinalities of finite sets
- How to order all the subsets of a given set using bit vectors
- The inclusion/exclusion principle and its applications
- The Cartesian product of any number of sets and the general distributive law for sets
- How to list and delist permutations

1. Let $A = \{1, 2, 3, 4\}$, $B = \{a, b, c, d, e\}$.
   a) Compute $|\mathcal{P}(\mathcal{P}_2(A \times (B \times A)))|$.
   b) Compute $|\mathcal{P}_3(A \cap B)|$.
   c) Compute $\sum_{k=0}^{|A \cup B|}(-1)^k |\mathcal{P}_k(A \cup B)|$.

2. You have a problem that you want to solve by inclusion/exclusion. You define 8 sets, $A_1, \ldots A_8$. What is the set whose cardinality is your ultimate objective to find? How many sets do you have to count to achieve your goal via inclusion/exclusion?

3. What is the 48th subset of $\mathbb{D}$ in bit vector order, taking the usual ordering on the digits. ($\emptyset$ is the 0th subset, and $\{0\}$ is the 1st.)

4. Let $A = \{a, b, c, d, e, f, g\}$.
   a) Taking $\emptyset$ as the 0th subset of A and $\{a\}$ as the 1st, what the 101st subset of A in bit vector order.
   b) Compute $|A \cup \mathcal{P}_2(A)|$.

5. Consider the set of numbers $A = \{1, 2, \ldots, 100\}$. How many numbers in A are either even, evenly divisible by 5, or evenly divisible by 7?

6. How many 8 letter words on $\mathbb{A}$ either start with three vowels (a, e, i, o, u) or end with two vowels?

7. How many strings in $\mathbb{A}^{12}$ are either three repetitions of strings of 4, such as *afczafczafcz*, or four repetitions of strings of 3, such as *gjigjigjigji*?

8. How many strings in $\mathbb{A}^{12}$ are either 2 repetitions of strings of 6, such as *pwertypwerty*, or three repetitions of strings of 4, such as *grokgrokgrok*?

9. How many strings in $\mathbb{A}^{30}$ are either 2 repetitions of strings of 15, 3 repetitions of strings of 10, or 5 repetitions of strings of 6?

10. Consider the strings of length 5 on $\{a, b, c, d\}$. What are the 10 strings following *abbbb* in lexicographical order?

11. What is the 1888th element of the set $(\{\alpha, \beta, \gamma, \delta, \epsilon\})^{10}$ in lexicographic order, taking the Greek letters in alpha-beta-ical order. (The order above.)

12. In considering the permutations of $\mathbb{D}$ in lexicographic order, with 0123456789 the 0th, which permutation is $9! + 8! + 7! + 6! + 5! + 4! + 3! + 2! + 1!$? Which permutation is $9! + 8! + 7! + 6! + 5! + 4! + 3! + 2! + 1! + 0!$?

Chapter 4

Formal logic

4.1 Statements and truth value

The study of logic begins well into the text because, at least in some sense, we are all familiar with its informal application. Logic is the use of careful reasoning applied to *statements*. Considering statements in terms of grammar, you might conclude that statements are very common, with any sentence being classified as either a question or a statement. But most non-questions are not statements in the sense of logic. For logic, a *statement is either true or false*.

Sherlock Holmes applies logic to solve crimes. He reasons that Sir Albert is the only man on the island who is left-handed, has knowledge of firearms, and knew where the rubies were hidden, and therefore, Sir Albert is the murderer. But Sherlock Holmes has a huge advantage over the police; not because he is well-educated and incredibly smart; but because he is a fictional character. Holmes inhabits a world of rigid facts created by an author. In the real world, there are right-handed people who sometimes use their left hand, there are war veterans who can't or won't use firearms, and there are men who look right at a heap of rubies hidden in a drawer but just don't notice them.

In the real world, treating facts as "facts" is a dangerous game which can let you build a structure as fragile as a house of cards. In detective fiction, the pleasure is often in the description of a long sequence of logical card houses built by minor characters and then blown down by new "clues", until the final structure is constructed by the protagonist. And that final one is only not blown away because the book is over.

The simplest assertion in natural language can have hidden problems which make it problematical. "Puerto Rico is a state" – True or False? What are we to say? It is not one of the fifty states of the United States, but it is a state in the sense of being a governmental unit. Right away we have two different reasonable interpretations. The problem is not that it is questionable whether the sentence is true or not. The problem is that it is *not clear what is meant*. Without context, it is not a logical statement at all.

It is much easier to find statements on which to apply logic in mathematics, games, or the law – areas where we humans make the rules and control the language which is used to express them.

- In chess, the bishop moves diagonally. (TRUE)
- $\pi > 3$. (TRUE)
- $\sqrt{2}$ is rational. (FALSE)

Discrete Mathematics With Logic. https://doi.org/10.1016/B978-0-44-318782-7.00009-5

- `<listy>`, `</listy>` are list item delimiters in HTML. (FALSE)

The third statement was concluded over two millennia ago, but the logical structure establishing it still stands because the foundational statements are firm.

In *formal logic* we study how logical statements interact. The *statements*, also called *propositions*, are indicated by variables, $p, q, r, \ldots$, and each has a *truth value* of TRUE, or FALSE.

It often helps to think of our logical variables as being replaced by a simple statement in natural language, e.g., r: "It is raining", p: "Paul is smart", but beware of slippery language and ambiguity.

More recently logic was given a more mathematical look by associating the truth values TRUE and FALSE with the bits 1 and 0 respectively, writing $p = 0$ instead of p is FALSE. In this formulation, p is called a *Boolean variable*, and 0 and 1 are called *Boolean values*.

One advantage of the Boolean formation is that it places TRUE and FALSE more on an equal footing. It is natural for us to love truth, and hate falsehood – to collect and cherish true statements and discard and scorn false ones. But that is not the point of view in logic. In logic, false statements are just as valuable as true ones. You can reason equally well from the knowledge that "7 is prime" is true, as from knowing that "6 is prime" is false.

Exercises

1. For each of the following, argue whether or not it qualifies as a statement in the sense of logic.
   a) Gas burns.   b) Canada is a peaceful nation.
   c) You can get anything you want at Alice's restaurant.
2. For each of the following, argue whether or not it qualifies as a statement in the sense of logic.
   a) The derivative of x^5 is $5x^4$.   b) Every rational number is a real number.
   c) $\lfloor \pi^{(2^{100})} \rfloor$ is prime.
3. Suppose that p, q, and r are statements. The expression "p and q or r" is ambiguous. Explain why.
   It may help if you let p stand for "Eddie has a penny", q stand for "Eddie has a quarter", and r stand for "Eddie has a ruby".

4.2   Logical operations

The statements of formal logic, the *Boolean variables* taking values 0 and 1, are not just considered in isolation. There are many *logical operators* connecting them, the most important being AND, OR, and NOT, symbolically $\wedge$, $\vee$, and $\neg$.

Definition 4.1. Let p and q be statements, that is, Boolean variables.
   The statement "p AND q", equivalently $p \wedge q$, asserts that both p and q are TRUE, i.e., have Boolean value 1.

The statement "p OR q", equivalently $p \vee q$, asserts that either p or q is TRUE, i.e., has Boolean value 1.

The statement "NOT p, equivalently $\neg p$, asserts that p is false, i.e., has Boolean value 0. ♠

In the definition of OR, it must be emphasized that we are using "or" inclusively, the default in this book, so the OR statement asserts that one, or the other, or both are TRUE.

With these operators, we can now form *Boolean expressions*, like $(p \wedge q) \vee (\neg p \wedge q \wedge \neg r)$, creating an algebra of logical propositions, or a *Boolean algebra*. It is useful to compare this Boolean algebra to the algebra you learned in high school for real numbers.

	Real Algebra	Boolean Algebra
Values	$0, -2/3, \pi$, etc.	0, 1
Variables	$x, y, z, \ldots$	$p, q, r, \ldots$
Operations	$+, \times, -$	$\wedge, \vee, \neg$
Expression	$r \times t = d$	$r \wedge (h \vee u)$
Natural World	Rate $\times$ Time $=$ Distance	It is raining and I need a hat or an umbrella

Basic Boolean identities

Here are some basic identities which govern Boolean algebra and connect the operators.

$$\neg(\neg p) = p \qquad \text{double negation}$$

$$
\begin{aligned}
p \vee \neg p = 1 \quad & p \wedge \neg p = 0 \\
p \vee 0 = p \quad & p \vee 1 = 1 \qquad \text{unit laws} \\
p \wedge 0 = 0 \quad & p \wedge 1 = p
\end{aligned}
$$

$$p \wedge p = p \vee p = p \qquad \text{idempotence} \qquad (4.1)$$

$$p \wedge q = q \wedge p; \quad p \wedge q = q \wedge p \qquad \text{commutativity}$$

$$p \wedge (q \wedge r) = (p \wedge q) \wedge r \quad p \vee (q \vee r) = (p \vee q) \vee r \qquad \text{associativity}$$

$$p \wedge (q \vee r) = (p \wedge q) \vee (p \wedge r) \qquad \text{Distributive law}$$

$$p \vee (q \wedge r) = (p \vee q) \wedge (p \vee r) \qquad \text{Distributive law}$$

$$\neg(p \vee q) = \neg p \wedge \neg q \qquad \text{De Morgan's law}$$

$$\neg(p \wedge q) = \neg p \vee \neg q \qquad \text{De Morgan's law}$$

In particular, just as with $\cap$ and $\cup$, each operator distributes over the other, making them more on an equal footing than $+$ and $\times$. In terms of order of operations, the negation $\neg$ gets the highest preference, just as $-$, so $\neg p \vee r$ is read as $(\neg p) \vee r$, not $\neg(p \vee r)$. For the binary operators $\wedge$ and $\vee$, however, we will make no preference, so $p \vee q \wedge r$ is regarded as an ill-formed expression, since $(p \vee q) \wedge r$ and $p \vee (q \wedge r)$ are not equal.

Just as in numerical algebra, one encounters problems where one searches for values of variables which make an expression true. Unlike real algebra, in Boolean algebra there are only two possible values per variable, so for N variables, it is possible to simply check all 2^N possibilities. For small examples this may be organized in a "truth table", but this is tedious and impractical in general.

Example 4.2. Find a truth assignment which makes each of the following Boolean expressions true.

$$(p \wedge \neg q \wedge r) \vee (q \wedge p \wedge \neg r) \vee (\neg p \wedge \neg q)$$
$$(p \vee \neg q \vee r) \wedge (q \vee p \vee \neg r) \wedge (\neg p \vee \neg q)$$

✠

First we should notice that the upper one is much easier to parse. Each term, or *clause*, is connected by an OR, so each clause may be examined separately. Since $p = r = 1$ and $q = 0$ makes the first clause TRUE, that assignment makes the whole expression TRUE.

For the second one, all three clauses must be TRUE so none can be ignored. Each clause is an OR statement, so can be satisfied in different ways. But we do not have to resort to checking all possibilities. Instead we look for commonalities and notice that p is an alternative for the first two terms, so taking $p = 1$ makes both of those terms TRUE, and leaves freedom in assigning q for the third term. Assigning $p = 1$ and $q = 0$ works independently of the truth value of r. ◇

Exercises

1. Suppose p and q are TRUE and r is false. Determine the truth value of each of the following:
   a) $(p \vee \neg q) \wedge (\neg q \vee r)$ b) $(p \wedge \neg q) \vee (\neg q \wedge r)$
2. Suppose $p = 0$, $q = 0$, $r = 1$, and $s = 0$. Find the Boolean value of each of the following:
   a) $(p \wedge \neg s) \vee (\neg q \wedge r)$ b) $(p \vee \neg s) \wedge (\neg q \vee r)$
3. Find Boolean values of each of p, q, r and s so that

$$(p \vee \neg q) \wedge (q \vee \neg r) \wedge (r \vee \neg s) \wedge (s \vee \neg p)$$

is true. Can you find a second assignment that works?

4.3 Implications

The logical operations that form the building blocks of Boolean algebra are $\vee$, $\wedge$, and $\neg$. There are several other minor operations which we will not concern ourselves with. But there is one other operation which we must learn: the *implication*. It is the fundamental underpinning of logic as it is used in arguments and reasoning, and it must be carefully studied to avoid common mistakes. Its definition in logic often comes as a surprise.

Definition 4.3. The *implication*, symbolically $\Rightarrow$, is a binary logical operation. The statement $p \Rightarrow q$ asserts that q is true or p is false:

$$(p \Rightarrow q) = (q \vee \neg p). \quad \spadesuit$$

In the expression $p \Rightarrow q$, p is referred to as the *antecedent*, or the *hypothesis*, and q is termed the *consequence* or the *conclusion*. So the definition of the implication above can be stated:

The consequence is true, or the antecedent is false.

In text, the implication $p \Rightarrow q$ can be variously expressed as, "p implies q", "if p then q", or "q is a consequence of p".

Beware that implications, are often not used the same way in logic as in natural language, so there are many pitfalls in understanding their meaning. In particular

- **No Temporality:** Unless time is part of the statement, the conclusion is not thought of becoming true *after* the antecedent is true, as is meant by saying "If you will just read this book then you will be rich."
- **No Causality:** In logic, the implication does not require that the truth of the antecedent causes the truth of the consequence. The connection is in truth value alone. In natural language, if you say "If you don't pay your taxes then you go to jail", the jail term is understood to be a consequence of tax avoidance.

Temporality is usually not a confusion in mathematical statements, but causality definitely is a common source of error.

$$(\pi > 3) \Rightarrow (3 \text{ is prime})$$

is a perfectly valid implication, since the consequence is true, but one is tempted to reject it as false because there is no obvious causal link.

This is even more pronounced in the equally valid

$$(27 \text{ is prime}) \Rightarrow (3 \text{ is prime})$$

which logic requires us to accept before we even read the consequence, since the hypothesis is false; and the frankly disturbing

$$(27 \text{ is prime}) \Rightarrow (1 + 1 = 7)$$

which is valid for the same reason.

Since the implication is an OR statement, it can be satisfied in many ways. The only way an implication can be false is if the hypothesis is true, and the consequence is false. This oddity in definition and meaning is not the result of the implication being flawed or old fashioned, it is source of its power. To understand that, it is necessary to distinguish how the implication is established, and how it is then used.

How to establish the implication $p \Rightarrow q$

First assume p to be true, then show, under that assumption, that q is true.

In showing the consequence, you may use the assumed truth of the antecedent, but you need not. Also, you are free to use anything else not specified in the implication, as long as it is true. After the consequence is concluded and the implication is established, the assumption that p be true is withdrawn, as well as the consequent conclusion that q is true. All that has been *proved* is $p \Rightarrow q$, the *implication itself.*

This establishment method works because of the nature of the OR statement $q \lor \neg p$. The reason you are allowed to assume p to be true is that, if p were false the OR would be satisfied, so p being true is the only case left to check.

Example 4.4. Suppose we wanted to prove that

If Empire apples are deadly poison, then you should not eat them.

You would start your argument by asking the reader to grant you the truth of the hypothesis for the sake of the argument. "Let Empire apples be deadly poison", you say, to the consternation of the apple industry. Once the hypothesis is granted, the argument to the conclusion is very easy, since you should not eat anything which is deadly poison, whatever it is. Now that the implication is established, our knowledge of the nature of an Empire apple returns to its former state, both with regard to it being poison, and the injunction not to eat it. Only the truth of the implication remains.   ◇

How to use the implication $p \Rightarrow q$

Show p. Then q follows immediately from the truth of $p \Rightarrow q$.

So, to use the implication, you do work to show logically that the antecedent is actually true. In that work, you need not concern yourself with q at all. If you are successful, and p is shown to be true, then the implication $p \Rightarrow q$ establishes that q is also true. It is equally valid, and maybe more convenient, to assume q is false and under that assumption show that p is false.

What makes this formulation so valuable is this division of labor. On the one hand, in the establishment, you show q without concern as to whether or why p should be true, and then, in the use, you are allowed to concentrate only on p without worrying about q at all. This is the reason that mathematicians, and those who use mathematics, are careful to preserve this ancient and rigid formulation of the implication despite natural language's drift in giving it other connotations. The implication in its pure form is too practically useful to alter.

Exercises

1. Which of the following implications do you regard as true?
   a) If it snowed in Boston last year then there will be a full moon there sometime next August.
   b) If you eat cauliflower every day then smoking cigarettes is a bad idea.
   c) If a leprechaun appears on St. Patrick's Day and gives you a pot of gold then U.S. federal law requires you to give all that gold to your parents.
2. Establish the following implications.
   a) If 18 is odd, then 20 is odd.
   b) If π is irrational, then 20 is even.
3. Establish the following implications:
   a) $(p \vee \neg p) \Rightarrow (q \vee \neg q)$
   b) $(p \wedge q) \Rightarrow (q \vee r)$

4.4   Double implication

If $p \Rightarrow q$ and $q \Rightarrow p$ are both true, sometimes written $p \Leftrightarrow q$, then p and q must have the same truth value. Here is why. If either p or q is true, one or the other implication shows that the other is true is well, so both are true. The only situation left, which must not be neglected, is if both p and q are false. By the definition of the implication, $p \Rightarrow q$ and $q \Rightarrow p$ are both true in this case as well, since for both the antecedent is false.

So $p \Leftrightarrow q$ says exactly that either $p = q = 0$ or $p = q = 1$. For this reason, the $\Leftrightarrow$ symbol is often used in logic in place of the equality symbol $=$. We will use both symbols synonymously:

$$[p = q] = [p \Leftrightarrow q] = [(p \Rightarrow q) \wedge (q \Rightarrow p)]$$

This identity justifies a technique that, like the double inclusion we studied earlier, is widely used, fundamental, and gives good practice in logical thinking and using logical notation. In this text it is called the *double implication method* of showing logical equivalence.

Definition 4.5. To establish the equivalence of two logical expressions p and q by the double implication method, show separately that $p \Rightarrow q$ and $q \Rightarrow p$.
♠

As a first application of the double implication method, we will show one of the distributive laws,

$$[p \wedge (q \vee r)] \Leftrightarrow [(p \wedge q) \vee (p \wedge r)]$$

to be followed by a short commentary.

Proof. First we show $[p \wedge (q \vee r)] \Rightarrow [(p \wedge q) \vee (p \wedge r)]$, so we assume $p \wedge (q \vee r)$ is true. Thus p and $q \vee r$ are both true. Since $q \vee r$ is true, there are two cases.

    Case 1. q is true. In that case $p \wedge q$ is true. Therefore $(p \wedge q) \vee (p \wedge r)$ is true.

    Case 2. r is true. In that case $p \wedge r$ is true. Therefore $(p \wedge q) \vee (p \wedge r)$ is true.

    So in either case, $(p \wedge q) \vee (p \wedge r)$, as required.

    Next we show $[(p \wedge q) \vee (p \wedge q)] \Rightarrow [p \wedge (q \vee r)]$. Let $(p \wedge q) \vee (p \wedge q)$ be true. Then there are two cases.

    Case 1. $p \wedge q$ is true, so both p and q are true. Since q, we have $q \vee r$, hence $p \wedge (q \vee r)$.

    Case 2. $p \wedge r$ is true, so both p and r are true. Since r, we have $q \vee r$, hence $p \wedge (q \vee r)$ in this case as well.

    Thus, in either case, $p \wedge (q \vee r)$, as required.

    Therefore the identity is true.   □

Commentary: The pattern is in the same form as the double inclusion proof, but the language and notation are different. Specifically, the requirement at each stage is different. Notice that to assert p, is the same as saying p is true, and, though not appearing in the argument above, to assert $\neg p$, is to say p is false. You could rewrite the whole argument without the words true and false, and substitute their bit equivalents 1 and 0.

Like with double inclusion, you should learn to write proofs of this form, and the best practice is using basic logical identities. But keep in mind, we are not concerned with reproving well-known identities. It is implication, and double implication that we are practicing.

Inverse, converse, and contrapositive

Given an implication $p \Rightarrow q$, the implication you would want to pair with it in the double implication method $q \Rightarrow p$ is called the *converse* of $p \Rightarrow q$. There are two others which have names:

- $q \Rightarrow p$ is the *converse* of $p \Rightarrow q$.
- $\neg p \Rightarrow \neg q$ is the *inverse* of $p \Rightarrow q$.
- $\neg q \Rightarrow \neg p$ is the *contrapositive* of $p \Rightarrow q$.

We have already seen that an implication and its converse are logically distinct, the two halves of the double implication method. However, every implication is logically equivalent to its contrapositive, and that implies that the inverse and converse are also logically equivalent *to one another*.

$$[p \Rightarrow q] \Leftrightarrow [\neg q \Rightarrow \neg p] \qquad [q \Rightarrow p] \Leftrightarrow [\neg p \Rightarrow \neg q]. \qquad (4.2)$$

Eq. (4.2) can quickly be shown from Definition 4.3 and Boolean algebra. Instead, it is instructive to prove $[p \Rightarrow q] \iff [\neg q \Rightarrow \neg p]$ as another example of the double implication method.

Proof. We will show first that $[p \Rightarrow q] \Rightarrow [\neg q \Rightarrow \neg p]$, so suppose that $p \Rightarrow q$. We now have to show the implication $\neg q \Rightarrow \neg p$ so let $\neg q$ be true, so q is false. Our objective is to show $\neg p$, that is, p is false. If p were true, then since $p \Rightarrow q$, we would have q, but q is false. Therefore p must be false, i.e., $\neg p$. We have shown $\neg q \Rightarrow \neg p$, $\heartsuit$ concluding the proof that $[p \Rightarrow q] \Rightarrow [\neg q \Rightarrow \neg p]$. $\diamondsuit$

The other half is left as an exercise. $\square$

Commentary. The argument is tricky because, inside the proof of one implication you have to prove a different implication, and that in different places in the argument you have to correctly establish and use implications. Notice what is known and what hypotheticals have been granted at $\heartsuit$. The implication $\neg q \Rightarrow \neg p$ has been shown, so the assumption $\neg q$ and the consequence $\neg p$ are no longer to be granted in the argument, as well as anything concluded from them *except* $\neg q \Rightarrow \neg p$. The overall implication $[p \Rightarrow q] \Rightarrow [\neg q \Rightarrow \neg p]$ is still being argued, so at $\heartsuit$ the hypothesis $p \Rightarrow q$ is still granted to be true. Fortunately, we have just what we need to establish the desired implication, and once that is done, at $\diamondsuit$, the first half of the argument is over, and the granted hypothetical $p \Rightarrow q$ and the concluded $\neg q \Rightarrow \neg p$ are released. All that remains proven is $[p \Rightarrow q] \Rightarrow [\neg q \Rightarrow \neg p]$.

Exercises

1. Prove by double inclusion the associative law for $\wedge$, the AND operator:

$$[p \wedge (q \wedge r)] \Leftrightarrow [(p \wedge q) \wedge r].$$

2. Prove one of De Morgan's Laws for formal logic, or one of the distributive laws, see Eq. (4.1), by the double implication method.

3. Consider the implication $a \wedge (b \vee c) \Rightarrow d$.
   a) Which of a, b, c, and d can you immediately assume to be true?
   b) What is the consequence of the implication?
   c) What is the antecedent of the contrapositive?
   d) What is the hypothesis of the converse?
   e) What is the conclusion of the inverse?

4.5 Working with Boolean algebra

The algebra of Boolean expressions is based on $\vee$, $\wedge$, and $\neg$ and follows the commutative, associative, distributive laws, and De Morgan's laws, and other equations listed in Eq. (4.1). Boolean algebra is the most direct way of working with complex logical expressions.

Example 4.6. Is the negation of the implication $p \Rightarrow q$ its inverse, its converse, or its contrapositive?

Actually, none of these, since $\neg(p \Rightarrow q) = \neg(q \vee \neg p) = \neg q \wedge \neg(\neg p) = p \wedge \neg q$, so the negation of an implication is not an OR statement at all, and is a stronger AND statement. $\diamond$

For a more complicated illustration of the algebraic approach, we will examine a Boolean identity once with double implication, and once with Boolean algebra:

$$[p \Rightarrow (q \Rightarrow r)] \Longleftrightarrow [(p \wedge q) \Rightarrow r]$$

Proof. (by double implication) We first show $[p \Rightarrow (q \Rightarrow r)] \Longrightarrow [(p \wedge q) \Rightarrow r]$. Assume $p \Rightarrow (q \Rightarrow r)$. We need to show $(p \wedge q) \Rightarrow r$ and so assume $p \wedge q$. So p and q are both true, and since $p \Rightarrow (q \Rightarrow r)$ we have $q \Rightarrow r$. Now, since q is true, we have r, as required to establish $(p \wedge q) \Rightarrow r$.

We next show $[(p \wedge q) \Rightarrow r] \Longrightarrow [p \Rightarrow (q \Rightarrow r)]$. Assume $(p \wedge q) \Rightarrow r$. To show $p \Rightarrow (q \Rightarrow r)$ we may assume p and have to show $q \Rightarrow r$. To show $q \Rightarrow r$ we may assume q. $\heartsuit$ Since we have p and q, we have $p \wedge q$ and $(p \wedge q) \Rightarrow r$ allows us to conclude r. Since assuming q allowed us to conclude r, we have $q \Rightarrow r$, and that conclusion allows us to say that $p \Rightarrow (q \Rightarrow r)$ as required in proving $[(p \wedge q) \Rightarrow r] \Longleftrightarrow [p \Rightarrow (q \Rightarrow r)]$. $\square$

Commentary. In the second part, count how many implications and hypotheticals we are juggling at $\heartsuit$. This is certainly the most complicated double implication proof one would ever want to read. But remember, we are not studying double implication to play with identities, but to learn the reasoning method to use later on other discrete problems. If you can get through this example, you can really say that you have mastered that method, and understand how implications work.

Here is the same expression approached via Boolean algebra, which is designed exactly to fit this type of problem.

$$
\begin{aligned}
[p \Rightarrow (q \Rightarrow r)] &= [(q \Rightarrow r) \vee \neg p] \\
&= [(r \vee \neg q) \vee \neg p] \\
&= [r \vee (\neg q \vee \neg p)] \\
&= [r \vee \neg(q \wedge p)] \\
&= [r \vee \neg(p \wedge q)] \\
&= [(p \wedge q) \Rightarrow r]
\end{aligned}
$$

You should definitely justify each step in the "calculation" above from the list in Eq. (4.1), and note the name of the property used.

Exercises

1. Express $p \Rightarrow (q \Rightarrow (r \Rightarrow s))$ only in terms of $\vee$, $\wedge$, and $\neg$.
2. Express $p = (q \Rightarrow r)$ only in terms of $\vee$, $\wedge$, and $\neg$.
3. Express $(p \Rightarrow q) \neq (q \Rightarrow p)$ only in terms of $\vee$, $\wedge$, and $\neg$.

4.6   Boolean functions

If we pursue the analogy between ordinary algebra and Boolean algebra, the next object to consider is the function. In ordinary algebra, you considered functions like $f(x) = 1 - 2x^2$, $g(x, y) = x^2 + y^2$, or $f(x, y, z) = (x - z)(y - x)$, you looked at their graphs, and studied their behavior. For each assignment of values to the independent variables, these functions determine the output value, which for these examples is just a single real number.

We can create the same structure in Boolean algebra and, in fact, we have already done so. Each Boolean expression we have examined, like $p \Rightarrow (q \Rightarrow r)$ determines for each set of Boolean values assigned to p, q, and r, an output Boolean value of 0 or 1, that is, TRUE or FALSE. We can make the situation look more like the algebra with which you are more familiar by writing $b(p, q, r) = [p \Rightarrow (q \Rightarrow r)]$, and noting that $b(0, 0, 0) = b(1, 1, 1) = 1$, and $b(1, 1, 0) = 0$. In logic, however, is it not very common to use the standard mathematical notation $b(p, q, r)$ to specify a Boolean function, but to simply note that an expression like $p \vee q \wedge (r \vee \neg p)$ defines a Boolean function.

Definition 4.7. A *Boolean function* on the Boolean variables p, q, r, ..., is a rule which assigns a bit, or TRUE or FALSE, to each assignment of bits to the variables $p, q, r, \ldots$.   ♠

Since there are 2^n possible ways to assign bits to n Boolean variables, the multiplicative principle gives us the following.

Theorem 4.8. *There are $2^{(2^n)}$ different Boolean functions on n Boolean variables, $p_1, \ldots p_n$.*

That's a lot of functions. So far, we have been specifying a Boolean function by choosing Boolean variables and forming an expression using any of the logical operators which we have defined, $\vee$, $\wedge$, and $\neg$. (We don't need $\Rightarrow$ since the implication can be always rewritten with $\vee$ and $\neg$.) So it would seem that $\neg$, $\vee$, and $\wedge$ will need to be augmented, as is done in ordinary functions when you added exotic functions like $\ln(x)$ and $\sin(x)$, to express them all. But for Boolean functions, no exotic functions are required.

Theorem 4.9. *Every Boolean function corresponds to a Boolean expression in the Boolean variables using only the operations $\vee$, $\wedge$, and $\neg$.*

We will illustrate the result for four Boolean variables, p, q, r, and s. Defining the function requires us to assign an output bit to each bit assignment of

p, q, r, and s. Suppose first, for a simple example, that we wanted to assign bit 1 in the situation $p = 1$, $q = 0$, $r = 1$, and $s = 1$, and assign 0 to all the other $2^4 - 1$ input assignments. That is easily accomplished by the AND clause $p \wedge \neg q \wedge r \wedge s$, and the same clause would assign FALSE to any other particular bit assignment to p, q, r, and s. If instead you had several input assignments to return TRUE, you can connect the individual AND clauses with $\vee$. With this trick you may express any of the $2^{(2^4)}$ Boolean functions. We can do a numerical check: there are 2^4 possible AND clauses, and for each of them, one chooses whether or not to include it in the final OR clause, so we compute $2^{(2^4)}$ different expressions of that type.

Example 4.10. Express a Boolean function on p, q, r, s and t which is TRUE when exactly one of the input variables is FALSE.

Assigning TRUE if just p is false corresponds to the AND clause $\neg p \wedge q \wedge r \wedge s \wedge t$, and the function can be expressed by

$$(\neg p \wedge q \wedge r \wedge s \wedge t) \vee (p \wedge \neg q \wedge r \wedge s \wedge t) \vee (p \wedge q \wedge \neg r \wedge s \wedge t) \vee$$
$$(p \wedge q \wedge r \wedge \neg s \wedge t) \vee (p \wedge q \wedge r \wedge s \wedge \neg t) \qquad \square$$

$\diamond$

Exercises

1. For p, q, r and s, express "they can't all be true" in terms of $\vee$, $\wedge$, and $\neg$.
2. For p, q, r and s, express "all are true or none are true" in terms of $\vee$, $\wedge$, and $\neg$.
3. For p, q, r and s, express "any three imply the fourth" in terms of $\vee$, $\wedge$, and $\neg$.

4.7 DNF and CNF†

Disjunctive normal form

By Theorem 4.9 every Boolean function may be expressed in terms of Boolean variables and $\wedge$, $\vee$, and $\neg$, but in the explanation, we actually showed that this could always be done in a special form, called a normal form.

Definition 4.11. An *AND clause* on the Boolean variables $p_1, \ldots p_N$ is a Boolean expression of the form $x_1 \wedge x_2 \wedge \cdots \wedge x_n$, where each x_i is either p_i or its negation $x_i \in \{p_i, \neg p_i\}$.

A Boolean expression is said to be in *disjunctive normal form* (DNF) if it can be written $A_1 \vee A_2 \vee \cdots \vee A_K$, where each A_i is an AND clause. ♠

The remarks following Theorem 4.9 actually show that

Theorem 4.12. *Every Boolean function can be expressed in disjunctive normal form.*

An expression in disjunctive normal form is slightly more general than the types thought of so far, in which each of the AND clauses had the same number of variables. The Boolean function $(p \wedge q) \vee (p \wedge \neg q \wedge r) \vee (p \wedge \neg q \wedge \neg r \wedge s)$ is also in disjunctive normal form. Also $(p \wedge \neg q \wedge r) \vee (s \wedge \neg t) \vee u$ is in disjunctive normal form, and notice the very short AND clause u at the very end.

Conjunctive normal form

We can consider the situation with AND and OR reversed.

Definition 4.13. An *OR clause* on the Boolean variables $p_1, \ldots p_N$ is a Boolean expression of the form $x_1 \vee x_2 \vee \cdots \vee x_n$, where each x_i is either p_i or its negation, $x_i \in \{p_i, \neg p_i\}$.

A Boolean expression is said to be in *conjunctive normal form* (CNF) if it can be written $O_1 \wedge O_2 \wedge \cdots \wedge O_K$, where each O_i is an OR clause. ♠

If you have an expression in conjunctive normal form, like

$$(p \vee \neg q \vee s) \wedge (q \vee \neg t \vee z) \wedge (r \vee s \vee t) \wedge (\neg p \vee \neg r \vee t)$$

it can be quite difficult to find a Boolean assignment to the variables which makes the expression TRUE. And, more concerning, the problem seems to get exponentially worse, the longer the expression gets, even though checking whether any assignment does or does not yield true is very quick. On the other hand, it is not known if there is some subtle trick which would allow us to solve the problem in a simpler way than, essentially, checking all possible assignments. Problems of this type are now well-studied, and are called NP-complete.

Theorem 4.14. *Every Boolean function can be expressed in conjunctive normal form.*

This follows either from first expressing the negation in disjunctive normal form, and then negating that expression in disjunctive normal form using De Morgan's laws, thereby switching all the ANDs to ORs, all the ORs to ANDs, and reversing which variables are negated in each clause. A second way to see this, is start with an expression of the desired Boolean function in disjunctive normal form, and use the distributive law to "multiply it out", just like you would multiply out $(x + y + z)(x - y + w)(-x + w + t)$. You can do that in reverse to convert a conjunctive normal form into an, easy to solve, disjunctive normal form, but that doesn't solve the NP-completeness problem since, in general, by the multiplicative principle, multiplying out the algebraic expression is still a lot of work.

Example 4.15. Write $(p \wedge \neg q \wedge r) \vee (\neg p \wedge q) \vee (p \wedge s)$ in CNF.

Multiplying out, by the multiplicative principle, will involve $3 \cdot 2 \cdot 2 = 12$ OR clauses. Here they are arranged "lexicographically" to make sure we don't get lost: $(r \vee q \vee s) \wedge (r \vee q \vee p) \wedge (r \vee \neg p \vee s) \wedge (r \vee \neg p \vee p) \wedge (\neg q \vee q \vee s) \wedge (\neg q \vee q \vee p) \wedge (\neg q \vee \neg p \vee s) \wedge (\neg q \vee \neg p \vee p) \wedge (p \vee q \vee s) \wedge (p \vee q \vee p) \wedge (p \vee \neg p \vee s) \wedge (p \vee \neg p \vee p)$. It really seems like we are not done. The expression contains OR clauses like $r \vee \neg p \vee p$ which ought to be simplified, (how?) but it is in CNF. $\diamond$

Exercises

1. For p, q, r, and s, express "they can't all be true" in boolean notation.
   a. Rewrite in DNF.
   b. Rewrite in CNF.
2. For p, q, r, and s, express "all are true or none are true" in boolean notation.
   a. Rewrite in DNF.
   b. Rewrite in CNF.
3. For p, q, r, and s, express "any three imply the fourth" in boolean notation.
   a. Rewrite in DNF.
   b. Rewrite in CNF.

4.8 Case study: Classic logic puzzles

Logic puzzles have been published on websites and blogs since the internet was created, and before that were a regular feature in newspapers for over a century. I first saw these problems given below in a paperback puzzle book published in the 1970's. The oldest version I can find cites a philosophy text by Copi from the 1960's, but I doubt that is the original source – from the wording of the problems it seems likely that they are at least 30 years older than that. They must be good problems because they usually appear together with only trivial alterations. I have tried to give you the problems in their most original formulations. (For solutions, usually bad ones, you can just Google the text, but where is the fun in that?)

★★★

   Benno Torelli, genial host of Jamtrack's most exclusive supper club, was shot and killed by a racketeer because he fell behind in his protection payments.
   When questioned by the police, each of the men made three statements, two true and one false.
   Lefty: "I did not kill Torelli. I never owned a revolver. Spike did it."
   Red: "I did not kill Torelli. I never owned a revolver. The other guys are all passing the buck."

Dopey: "I am innocent. I never saw Butch before. Spike is guilty."

Spike: "I am innocent. Butch is the guilty man. Lefty lied when he said I did it."

Butch: "I did not kill Torelli. Red is the guilty man. Dopey and I are old pals."

Daniel Kilraine was killed on a lonely road, two miles from Pontiac, at 3:30am on March 17. Otto, Curly, Slim, Mickey, and The Kid were arrested a week later in Detroit and questioned.

Each of the five made four statements, three of which were true, and one of which was false.

One of these men killed Kilraine. The statements were:

Otto: "I was in Chicago when Kilraine was murdered. I never killed anyone. The Kid is the guilty man. Mickey and I are pals."

Curly: "I did not kill Kilraine. I never owned a revolver in my life. The Kid knows me. I was in Detroit the night of March 17."

Slim: "Curly lied when he said he never owned a revolver. The murder was committed on St. Patrick's day. Otto was in Chicago at this time. One of us is guilty."

Mickey: "I did not kill Kilraine. The Kid has never been in Pontiac. I never saw Otto before. Curly was in Detroit with me on the night of March 17."

The Kid: "I did not kill Kilraine. I have never been in Pontiac. I never saw Curly before. Otto lied when he said I am guilty."

The employees of a small loan company are Mr. Black, Mr. White, Mrs. Coffee, Miss Ambrose, Mr. Kelly, and Miss Earnshaw. The positions they occupy are manager, assistant manager, cashier, stenographer, teller, and clerk, though not necessarily in that order.

The assistant manager is the manager's grandson, the cashier is the stenographer's son-in-law, Mr. Black is a bachelor, Mr. White is twenty-two years old, Miss Ambrose is the teller's stepsister, and Mr. Kelly is the manager's neighbor.

Who holds each position?

4.9   Case study: Spies

Suppose you were designing a spy game. To make the game interesting, your spies can be equipped with a variety of unique gadgets. So far you have programmed in the following:

a	antigravity belt with signal flairs
b	brick of gold
c	car with $100K£$ in the glove box
d	dagger with diamond hilt
e	pearl earring with hidden transmitter
f	faithful sidekick
g	gun with silver bullets
h	helicopter with missile launcher

Each spy will not be given all these fancy gadgets, just a subset. But the spy must have enough gear to be survivable. To be survivable he must have among his gadgets a weapon, something valuable, a method of communication, and transport. $\{a, b, c, d, e, f, g, h\}$ are Boolean variables which are true if the spy possesses the corresponding gadget, and false otherwise.

Here is a proposed survivability function:

$$(b \vee d \vee g \vee h) \wedge (b \vee c \vee d \vee e \vee g) \wedge (a \vee e) \wedge (a \vee c \vee h)$$

Each of the four OR clauses has a meaning in the game. Can you identify which is which?

Each represents one of the four conditions of survivability, possession of weapon, valuables communicator, and transport respectively. So the brick of gold is a valuable which the spy, at necessity, could use as a weapon, but not for transport or for communication. Of course, the faithful sidekick doesn't help the spy survive at all, and doesn't occur in the survivability function.

The survivability function, which is naturally and easily written in CNF, illustrates that CNF is not nearly so useless as it seemed when considering the Boolean function only from the point of view of finding Boolean values which evaluate it to TRUE.

The negation, and a little Boolean algebra, quickly gets us an expression for the non-survivable function in DNF:

$$(\neg b \wedge \neg d \wedge \neg g \wedge \neg h) \vee (\neg b \wedge \neg c \wedge \neg d \wedge \neg e \wedge \neg g) \vee (\neg a \wedge \neg e) \vee (\neg a \wedge \neg c \wedge \neg h)$$

Multiplying out to find the CNF of non-survivability will involve us in $4 \cdot 5 \cdot 2 \cdot 3$ terms, a large number but we expected that, which then would be culled for duplicates.

An interesting question would be, since the OR clauses of the survivability function in CNF, neatly expressed, have meaning in the game, why shouldn't the OR clauses of the CNF for non-survivability also have some meaning? Does that CNF hold other concepts of strategic value, perhaps known to us already, perhaps revealing a new insight?

4.10   Case study: Pirates and cannonballs

Pirates seem always to be popular, adventurously sailing the high seas, looting and plundering and being in every way perfectly admirable folk. Some people have favorite pirate heroes, fictional or actual, Long John Silver or Blackbeard. But asked who was the greatest sea pirate of them all, if you judge fairly you would certainly have to settle on a woman, Queen Elizabeth I of England. Her privateers were hugely successful, filled the treasury of England with heaps of gold, and included titled noblemen such as Sir Walter Raleigh, famous also for spreading his cloak in the mire.

Sir Walter, personally, was so successful as a pirate that he could afford such luxuries on his ship as a mathematician, who did work on the problem we are considering in this case study.

Suppose that you have plenty of gunpowder and cannonballs, all prepared to help persuade any ships you encounter to hand over their treasure. The cannonballs are neatly arranged in a stack with a square wooden frame containing $n \times n$ balls. In the chinks between the balls of the bottom layer, there are places for $(n - 1) \times (n - 1)$ balls in the next layer, and so on to the next layer and the next until a single cannonball is placed on top. *The question is, how many cannonballs are in the stack?*

The internet being intermittent aboard ship, the answer is difficult to look up, so they start making a table, recording first the different heights from 1 to 10, then the number of balls in each layer, and with a little trouble, they fill in the third column, the number of balls in each stack from 1 to 10. We can imagine Sir Walter and his employee looking for a pattern in values of column 3 of Table 4.1 in order to figure out the general formula. But, at least at first, they don't see one. Do you?

TABLE 4.1 Stacking cannonballs with a square base.

n	n^2	$1^2 + \cdots + n^2$	$\sum_{i=1}^{n} i^2$
1	1	1	1
2	4	5	5
3	9	14	$2 \cdot 7$
4	16	30	$2 \cdot 3 \cdot 5$
5	25	55	$5 \cdot 11$
6	36	91	$7 \cdot 13$
7	49	140	$2^2 \cdot 5 \cdot 7$
8	64	204	$2^2 \cdot 3 \cdot 17$
9	81	285	$3 \cdot 5 \cdot 19$
10	100	385	$5 \cdot 7 \cdot 11$

✠

After a while, in desperation, they factor each of the totals into their prime factors, placing the results in column 4, and are quite disappointed with the results. But there is a sort of pattern, intermittent but persistent, like a glimpsed shadow in the fog which might be a prize ship to be attacked. The pattern is underlined in column 4: skip the first entry, 5, 7, skip, 11, 13, skip, 17, 19, skip.

We would be happy to fill in the pattern even with non-primes to get 3, 5, 7, 9, 11, 13, 15, 17, 19, 21, but each of those skipped entries is missing a factor of 3. So, to make the pattern work, and at a loss for anything better to do, they add a factor of three to all entries, dividing it right back out again, so that the number is unaltered. The result is column 4 in Table 4.2, with the numbers in the desired pattern in parentheses.

TABLE 4.2 The pattern coming into focus.

n	n^2	$1^2 + \cdots + n^2$	$\sum_{i=1}^{n} i^2$	$\sum_{i=1}^{n} i^2$
1	1	1	$1 \cdot (3)/3$	$[1] \cdot \langle 1 \rangle \cdot (3)/3$
2	4	5	$(5) \cdot 3/3$	$[1] \cdot \langle 3 \rangle \cdot (5)/3$
3	9	14	$2 \cdot (7) \cdot 3/3$	$[2] \cdot \langle 3 \rangle \cdot (7)/3$
4	16	30	$2 \cdot (9) \cdot 5/3$	$[2] \cdot \langle 5 \rangle \cdot (9)/3$
5	25	55	$5 \cdot (11) \cdot 3/3$	$[3] \cdot \langle 5 \rangle \cdot (11)/3$
6	36	91	$7 \cdot (13) \cdot 3/3$	$[3] \cdot \langle 7 \rangle \cdot (13)/3$
7	49	140	$2^2 \cdot 7 \cdot (15)/3$	$[4] \cdot \langle 7 \rangle \cdot (15)/3$
8	64	204	$2^2 \cdot 3 \cdot (17) \cdot 3/3$	$[4] \cdot \langle 9 \rangle \cdot (17)3$
9	81	285	$3 \cdot 5 \cdot (19) \cdot 3/3$	$[5] \cdot \langle 9 \rangle \cdot (19)/3$
10	100	385	$5 \cdot 11 \cdot (21)/3$	$[5] \cdot \langle 11 \rangle \cdot (21)/3$

Looking for a pattern in the numbers not in parentheses is still quite disappointing. But something is there, again not a regularly ascending or descending sequence, but one hobbling along in twos – from bottom to top you see 5, 5, 4, 4, 3, 3, 2, 2, skip 1. The pirates add the missing factor of 1 at no cost and isolate those numbers in square brackets. That takes them to column 5, with the leftover numbers in angular brackets. In the angular brackets we have another hobbling pattern, bottom to top 11, 9, 9, 7, 7, 5, 5, 3, 3, 1, but this is actually progress. Why not write down separate formulas for the even and odd cases? For the even height stacks they quickly get $(n/2)(n + 1)(2n + 1)(1/3)$. For the odd height stacks get $n((n + 1)/2)(2n + 1)(1/3)$.

Break out the rum – algebraically it is *the same formula in both cases*!

The next thing they do is just what you or I would do. They add 11^2 to 385 to get $385 + 121 = 506$ for a stack of height 11. Then they compute

$$\frac{11(11 + 1)(2 \cdot 11 + 1)}{6} = 506$$

and find – hurrah! – that their formula works!

Hurrah! – but now they are completely stuck. They can check their formula against new cases, 12, 13, 14, ..., but that does not provide a convincing logical argument to prove the general formula. Their "successful" method started with the pattern of primes – for which they had no formula in 1600, and there is no formula even now. So those pirates, even with the help of logic, had no way scaling their method to show that their formula was true in the general case.

But that mathematician had a way. That will have to wait until the next chapter.

4.11   Summary exercises

You should have learned about:

- Statements and formal logic
- Proper notation in formal logic
- Formal logic operations, $\vee$, $\wedge$, and $\neg$
- Expressing formal logical statements in the Boolean algebra
- The implication $\Rightarrow$ in formal logic
- That an implication is equivalent to its contrapositive
- How to establish versus how to use an implication
- How to write a double implication proof

1. Discuss whether each of the following qualifies as a statement, and if so, whether it is true or false.
   a) In chess, the bishop moves diagonally.
   b) Napoleon Bonaparte is alive and hiding in the International Space Station.
   c) $\pi > 2$.
   d) π^2 is rational.
   e) π is wonderful.
   f) Pluto is a planet.
   g) SpaceX launches bomb because of snarky TV commentators.
2. Let p_0, p_1, p_2, and p_3 be Boolean variables. Express "at most 2 are true" as a Boolean expression.
3. For each of the following sentences, rewrite it as a standard implication and decide whether the implication is true in the logical sense: Say whether the sentence in natural language involves temporality, causality, or both.
   a) "If you can't stand the heat, get out of the kitchen."
   b) "If you are not part of the solution, then you are part of the problem."
   c) "Choosy mothers choose JIFF."
   d) "Even if you're on the right track, you'll get run over if you just sit there."
   e) "If you can read this, then you are too close."
4. Find two assignments of truth values for the variables p, q, and r which make $p \Rightarrow (q \vee \neg r)$ true.

5. If you wanted to prove directly the contrapositive of $(r \lor p) \Rightarrow (q \lor \neg r)$, what would you be allowed to assume?

6. What is the antecedent of the implication $p \Rightarrow (q \Rightarrow (r \lor p))$? What is the conclusion of the contrapositive? What is the hypothesis of the contrapositive?

7. Write the negation of $(r \lor p) \Rightarrow (q \lor \neg r)$ so that no parenthetical expression is negated.

8. Write the negation of $p \Rightarrow (q \Rightarrow (r \lor p))$ as a Boolean function so that no parenthetical expression is negated.

9. Find two Boolean assignments for p, q, and r for which $p \Rightarrow (q \lor \neg r)$ and $\neg p \lor \neg q \lor \neg r$ are both true.

10. Use the distributive law to write $(((a \land b) \lor c) \land d) \lor (d \land e)$ so that no symbol $\lor$ is inside any parenthesis.

11. Use the distributive law to write $(((a \land b) \lor c) \land d) \lor (d \land e)$ so that no symbol $\land$ is inside any parenthesis.

12. Let $B(p, q, r) = (1 - pq)^2(1 - (1 - q)(1 - pr))$. Verify that for p, q, $r \in \mathbb{B}$, that $B(p, q, r) \in \mathbb{B}$. Rewrite $B(p, q, r)$ using $\lor$, $\land$, and $\neg$.

Chapter 5

Induction

5.1 Predicate logic

Formal logic is based on statements, equivalently propositions, equivalently Boolean variables. Predicate logic is based on predicates. You may know the word predicate from grammar, the study of languages in which sentences are taken apart and their structure analyzed. A sentence may be decomposed grammatically into a *subject*, a noun phrase; followed by a *predicate*, consisting of most everything else – the verb, the objects, the prepositional phrases, and all the other parts of speech which come together in the sentence to express what you want to say about the subject.

"**Benedict Arnold** *is a filthy traitor and ought to be tarred and feathered.*" This is a simple sentence with the subject in bold and the predicate in italics. Logically the sentence might stand for a proposition which could be true or false depending on your opinion of Benedict Arnold.

In predicate logic, the idea is to hold the predicate fixed, and to vary the subject.

"**John Adams** *is a filthy traitor and ought to be tarred and feathered.*"
"**Wyatt Earp** *is a filthy traitor and ought to be tarred and feathered.*"
"**John Philip Sousa** *is a filthy traitor and ought to be tarred and feathered.*" This idea of predicate may be conveyed by using a generic unspecified subject, a pronoun indicating the set of people under consideration: "**The accused** *is a filthy traitor and ought to be tarred and feathered.*" The particular instances above are said to have been *instantiated*, or *quantified* from this general version.

Here is a more mathematical example, "**The number** *is a prime number*"; which may be quantified to be true, "**Twenty-three** *is a prime number*", or false "**Avogadro's number** *is a prime number*".

This linguistic game is how predicate logic got its name, but mathematically it is more efficient to regard the predicate as a function which assigns a truth value, that is, a Boolean value to each element of the set of subjects.

Definition 5.1. A *Boolean function* P on a set S is an assignment $P(s) \in \mathbb{B}$ for each element $s \in S$. ♠

Note, $P(s)$ is often written as P_s.

The Boolean functions we had in formal logic fit this scheme as a special case. For example for $r \vee (p \Rightarrow q)$ the elements of S are triples of Boolean values, and $r \vee (p \Rightarrow q)$ is TRUE if quantified with $r = p = q = 1$, and FALSE

Discrete Mathematics With Logic. https://doi.org/10.1016/B978-0-44-318782-7.00010-1

if instantiated with $r = p = 0$ and $q = 1$. The new, more general concept of Boolean function allows us to bridge the gap between the formal logic we have been studying, and sets we have been designing to help solve problems. With predicate logic, sets and logic can be made to work together and support one another.

When the predicate is described in the terminology and notation of Boolean functions, the Boolean function is usually said to be *evaluated*, rather than *instantiated* or *quantified*, although all three terms are acceptable.

Consider the set $\mathbb{N}$ of natural numbers, with $n \in \mathbb{N}$, and set $P(n) = [n \geq \pi]$. So $P(0) = P(1) = P(2) = P(3) = 0$, or FALSE, with all other values evaluating to 1 or TRUE.

Another example on the natural numbers is $Q(n) = [n^2 + 10 = 7n]$, for which $Q(2) = Q(5) = 1$, with all other natural numbers evaluating to FALSE. It is a little odd that the equals sign is occurring twice in two different senses in the definition of $Q(n)$. For this reason some prefer to use an alternate symbol, such as $Q(n) := [n^2 + 10 = 7n]$, which we will often follow. Note: it is a common mistake to evaluate $Q(5)$ as 35, since both sides of the equality evaluate as 35. Instead $Q(5)$ evaluates to 1 indicating that the equality $5^2 + 10 = 7 \cdot 5$ is true.

Here is another example on a different set, $\mathcal{P}(\mathbb{D})$. Suppose we want to consider whether or not subsets of $\mathbb{D}$ have even cardinality or not. $R(A) := [|A| \text{ is even}]$, So $R(\emptyset) = R(\{1, 2, 3, 4\}) = R(\mathbb{D}) = 1$, and $R(\{1, 3, 4\}) = 0$.

Example 5.2. With predicate logic, you may see even greater utility in the definition of the implication in logic: the consequence is true or the antecedent is false; since there are various instantiations to compare.

> **"The customer** *over six feet tall may not rent a rowboat."*

The rule is TRUE for Hans who is 6'8" and may not rent a rowboat (antecedent and consequence both true). It is true for Ignaz, who is 5'2" and may rent a rowboat (antecedent false and consequence false). It is valid also for Jörg, who is 5'1" but may not rent a rowboat because he weighs 288 lbs. (antecedent false and consequence true). The company violates the height rule if it rents a rowboat to Hans. It does not violate the height rule if it rents rowboats to Ignaz and Jörg, or refuses to do so. The point of this example is not to focus just on Hans, but on the totality of customers. The implication should be valid for all, but is written solely to prevent tall people from renting rowboats.   ◇

Exercises

1. Underline the predicate of each of the following, and instantiate each subject twice, once resulting in TRUE, once resulting in FALSE.
   a. The man was born after 1888.
   b. The state is north of North Carolina.

2. Underline the predicate of each of the following, and instantiate each subject twice, once resulting in TRUE, once resulting in FALSE. Context: Our universe are the natural numbers.

   a. The number is the sum of three distinct cubes.

   b. The number is the product of three primes.

3. The point lies between one and two units from either $(1, 1)$ or $(-1, -1)$.

   Draw a sketch of the points in the plane in which the points making this Boolean function true are colored red, and those making it false are colored blue.

5.2 Existential and universal quantification

The general Boolean functions we have introduced only become statements and have a truth value when they are evaluated. They have no truth value in and of themselves, just as $f(x) = x^2 + 1$ has no numerical value, while $f(42)$ does.

Boolean functions defined *on the same set* may be combined with our logical operators $\wedge$, $\vee$, $\neg$, and $\Rightarrow$. So, if we have defined $P(n) := [n^2 > 42]$ and $Q(n) := [n^3 \geq 1000]$, then $P(n) \wedge \neg Q(n)$ would evaluate to true for $n = 7$ and false for $n = 1$.

The logical operators of formal logic, when applied to Boolean functions, satisfy all the Boolean identities listed in Eq. (4.1). In particular, De Morgan's laws and the distributive laws hold. So there is an algebra of Boolean functions extending our formal Boolean algebra in just the same way that the ordinary algebra of numbers was extended to an algebra of real functions in high school.

	Algebra of real functions	Algebra of Boolean functions
Values	$0, -2/3, \pi$, etc.	$0, 1$
Variables	$x, y, z \in \mathbb{R}$	$s \in S$
Operations	$f(x) + g(x)$, $f(x)g(x)$, $-f(z)$	$P_s \wedge Q_s$, $P_s \vee \neg Q_s$

One difference between the algebra of real functions and general Boolean algebra is that predicate logic has two very important special forms of evaluation.

Definition 5.3. The *universal quantification* of the Boolean function P defined on the set S is the statement that $P(s)$ is true for all elements $s \in S$. ♠

There is a special logical symbol $\forall$, read as "for all", defined by

$$[\forall s \in S; P(s)] = \left[\bigwedge_{s \in S} P(s) \right]$$

which is used for universal quantification.

Notice that $\forall s \in S; P(s)$ is a simple statement – it is either true or false – and does not "depend on s". All predicates in the expression have been quantified already.

So $\forall n \in \mathbb{N};\, n^2 > 20$ is simply false, while $\forall n \in \mathbb{Q};\, n^2 \neq 2$ is true, since $\sqrt{2}$ is irrational. How about $\forall n \in \mathbb{N};\, n^3 - n$ is divisible by 3? Can you determine its truth value?

In contrast to universal quantification is *existential quantification*, which sounds very exotic and intimidating, but is just the OR version of the previous concept:

Definition 5.4. The *existential quantification* of the Boolean function P defined on the set S is the statement that there exists some $s \in S$ such that $P(s)$ is true. ♠

There is a special logical symbol $\exists$, read as "there exists", and defined by

$$[\exists s \in S;\, P(s)] = \left[\bigvee_{s \in S} P(s) \right]$$

which is used for existential quantification.

Again $\exists s \in S;\, P(s)$ is either true or false. $\exists n \in \mathbb{N};\, [n^2 + 4 = 4n]$ is true since $2^2 + 4 = 4 \cdot 2$, and $\exists n \in \mathbb{Z};\, [n^2 + 1 = 0]$ is false even if you believe in complex numbers, because $\sqrt{-1} \notin \mathbb{Z}$.

Establishing a universally quantified statement ($\forall s \in S;\, P_s$)

For a small finite set S you might be able to check P_s for every single element $s \in S$. Otherwise you have to show P_s for a generic s. You start *let $s \in S$ be given*, and then prove P_s, with the proof being valid *regardless* of which subject s was given. You may need cases, but other than that you may only use the properties accruing to s by the fact that $s \in S$. No further assumptions about s are allowed.

Using a universally quantified statement ($\forall s \in S;\, P(s)$)

Given the universally quantified statement, *you* may choose whichever $s \in S$ you prefer. The universally quantified statement guarantees that P_s is true. And if you pick a different subject, $s' \in S$, $P_{s'}$ is also true.

Establishing an existentially quantified statement ($\exists s \in S;\, P_s$)

Now *you* may choose any $s \in S$. You may choose a particular s you know well, or you may construct an s from ingredients you know exist, but don't know specifically, like the 30 trillionth prime number. There is no restriction on s beyond that it is an element of S. Then you only need to prove P_s for your chosen s. That done, the existentially quantified statement is true. (You might want to make a note that, for your chosen s, P_s is true, since you cannot recover that fact just from knowing $\exists s \in S;\, P_s$.)

Using an existentially quantified statement $(\exists s \in S; P_s)$

You may assume the existence of an $s \in S$ for which $P(s)$ is true. From $[\exists s \in S; P_s]$ you have absolutely no control over which s, or whether there is more than one. In particular, if you have already chosen $s \in S$, you may *not* conclude P_s is true because $[\exists s \in S; P_s]$ is true. Don't be thrown off because you are already considering an element of S actually called s. The letter s in $[\exists s \in S; P_s]$ is a "dummy variable", having no connection to the value of any variable outside the brackets of the statement. It should be clear that

$$[\exists s \in S; P_s] \Longleftrightarrow [\exists \varpi \in S; P_\varpi]$$

The universally and existentially quantified predicates are defined in terms of $\wedge$ and $\vee$, so they simply follow De Morgan's laws and the distributive laws.

$$\neg[\forall s \in S; P_s] = [\exists s \in S; \neg P_s], \quad \neg[\exists s \in S; P_s] = [\forall s \in S; \neg P_s]$$

Exercises

1. Decide whether or not the following is true or false, and prove it.

$$\exists n \in \mathbb{N}; n^2 + 5 = 6n$$

2. Decide whether or not the following is true or false. If it is true, then prove it.

$$\forall n \in \mathbb{N}; n^2 \text{ is even or } n^3 \text{ is odd}$$

3. State whether the expression is a statement or merely a predicate. If a statement, state whether or not it is true.
    a. $\forall n \in \mathbb{N}; \exists m \in \mathbb{N}; n^3 \geq m^2$
    b. $\exists m \in \mathbb{N}; \forall n \in \mathbb{N}; n^3 \geq m^2$
    c. $\exists n \in \mathbb{N}; \forall m \in \mathbb{N}; n^3 \geq m^2$
    d. $\forall m \in \mathbb{N}; \exists n \in \mathbb{N}; n^3 \geq m^2$
    e. $\forall n \in \mathbb{N}; \forall m \in \mathbb{N}; n^3 \geq m^2$
    f. $\exists n \in \mathbb{N}; \exists m \in \mathbb{N}; n^3 \geq m^2$

5.3　The theory of induction

In the previous section we noted that, among the four tasks associated with universally and existentially quantified predicates, the hardest one was to establish a universally quantified statement over a large set. In this section, we show how to prove the universal quantification of a predicate over the infinite set $\mathbb{N}$ using the method of induction.

Induction may seem tricky and difficult at first, but you should not be surprised if, after a time, it has become your favorite proof technique. For now, your objective as a student should be to understand how and why induction works theoretically; how to critically listen to inductive arguments and be able to spot any flaws if there are any; to be convinced by inductive arguments if they are valid; and even be able write simple induction proofs yourself.

Definition 5.5. A *proof by induction* is an argument that $[\forall n \in \mathbb{N};\ P_n]$ is true because the statement $[P_0 \wedge (\forall\ n \in \mathbb{N};\ P_n \Rightarrow P_{n+1})]$ is true.   ♠

   The induction procedure requires, instead of the proof of all instances of P_n, the establishment of the implications $P_n \Rightarrow P_{n+1}$. Theoretically, implications are weaker statements and hence "easier" to prove. Practically, you can only make use of that advantage if you understand how to establish implications.

   Some people prefer to see Eq. (5.1) in slightly different notation.

$$
\left[\bigwedge_{n=0}^{\infty} P_n \right] = [P_0 \wedge (\forall\ n \in \mathbb{N};\ P_n \Rightarrow P_{n+1})]
$$

$$
= [P_0 \wedge (P_0 \Rightarrow P_1) \wedge (P_1 \Rightarrow P_2) \wedge (P_2 \Rightarrow P_3) \wedge \cdots] \tag{5.1}
$$

$$
= P_0 \wedge \left[\bigwedge_{n=0}^{\infty} [P_n \Rightarrow P_{n+1}] \right]
$$

You may use the version you like best.

   Here is some helpful terminology which is commonly used. In the inductive formation $P_0 \wedge [\forall n \in \mathbb{N};\ (P_n \Rightarrow P_{n+1})]$, the proposition P_0 is called the *base case*. It must be shown separately from the other, more intimidating term. To prove $[\forall\ n \in \mathbb{N};\ P_n \Rightarrow P_{n+1}]$, as with any universally quantified statement, we start by letting a general $n \in \mathbb{N}$ be given, and then show, for that given n, the truth of the implication $P_n \Rightarrow P_{n+1}$. At that point, the implication $P_n \Rightarrow P_{n+1}$ is called *the induction step*, and the instance P_n, which you may assume to be true in its establishment, is called the *induction hypothesis*.

Proof of Eq. (5.1). This is to show why induction is a valid method. We use the double implication method. If the left hand side is true then all instances P_i are true, in which case P_0 is true and for any given n the implication $P_n \Rightarrow P_{n+1}$ is true because the consequence is true. (That was fast.)

   Now suppose the right hand side is true. So P_0 is true and every instance of the implication $P_n \Rightarrow P_{n+1}$ is true as well. Now we want to use those implications to show that all the individual instances of P_i are also true. Since P_0 is true, we know that at least some of the P_i's are true. If any are false, there must be a smallest number k, with P_k false. Since k is the smallest, and $k \neq 0$, P_{k-1} is true. But we have the implication $P_{k-1} \Rightarrow P_{(k-1)+1}$, which means that P_k cannot be false after all, a contradiction. Thus $[\forall n \in \mathbb{N};\ P_n]$ is true.   □

   If you don't like that formal approach, at least you notice that it is the second half of the double implication method, which is the difficult part. A more direct argument is: P_0 is true, and because $P_0 \Rightarrow P_1$, we conclude P_1 is true; but now, because $P_1 \Rightarrow P_2$ is true, we conclude P_2 is true too. But now, because $P_2 \Rightarrow P_3$ is true, we conclude ....

So eventually all the P_n are true.

We know there is no "eventually" in logic, but still the second version probably seems more convincing, even with the vague "..." to finish it off.

Example 5.6. Suppose you have [∀ $n \in \mathbb{N}$; $P_n \Rightarrow P_{n+1}$]. What can you conclude?

☩

In the first place, if P_0 is true, then that is the missing base case, and we conclude that all instances P_n are true by induction, [∀ $n \in \mathbb{N}$; P_n].

But what if P_0 is false? In fact, what if they are all false? Then by the properties of the implication, $P_n \Rightarrow P_{n+1}$ is TRUE. So both scenarios are possible, all instances if P_n true, and all P_n false.

Does that mean that anything can happen? No. If P_{77} is true, then the implications imply P_{78}, P_{79}, P_{80}, ... and so on for all values of $n \geq 77$. The number 77 acts as a base case for an inductive argument, and we conclude [∀ $n \geq 77$; P_n]. What about P_{76}? Could that be false? Actually, yes. The implication $P_{76} \Rightarrow P_{77}$ is true since the consequence is true, so both truth values for P_{76} are consistent with everything we have so far.

However, if we accept $\neg P_{76}$ there can be no $n < 76$ with P_n TRUE. If, say P_{23} were true, then P_{23} would act as the base case of an inductive argument, and P_n would be true for all $n \geq 23$, including $n = 76$. So if we allow P_{77} to be TRUE and P_{76} to be FALSE, then this is the one tipping point. P_n is TRUE for all $n \geq 77$, and P_n is FALSE for all $n \leq 76$.

So, given [∀ $n \in \mathbb{N}$; $P_n \Rightarrow P_{n+1}$], there are three types of scenarios possible for the individual instances: all instances true, all instances false, or a finite sequence of falsehood, followed by an endless sequence of truth. ◇

Exercises

1. Suppose you wanted to prove by induction that $3 \cdot n! > n^2$ for all $n \geq 0$.
   What is the base case?
   What implication would you have to show for the induction step?
   What induction hypothesis would you be allowed to assume to prove the induction step?

2. Suppose you wanted to prove by induction that, for every natural number, $(2n + 3)! \geq 10^{2(n+2)}$.
   What is the base case?
   What implication would you have to show for the induction step?
   What induction hypothesis would you be allowed to assume to prove the induction step?

3. Suppose P_n is a predicate defined on $n \in \mathbb{N}$. Suppose that $\forall n \in \mathbb{N}$; [$P_n \Rightarrow P_{n+1}$], and that $P_{21} \wedge \neg P_{12}$ is true. For each of the following, state whether

it must be true, must be false, or cannot be concluded from the information given: P_5, P_{15}, P_{25}, P_{35}, and P_{45}.

5.4   Induction practice

Now let's try out the method of induction on a real predicate of interest, and we should start with the equation which was left hanging from Section 4.10, the case study on pirates and cannonballs:

$$\sum_{k=0}^{n} k^2 = \frac{n(n+1)(2n+1)}{6}$$

If you read that, we left off with a formula which we strongly suspected was true for all n, but for which we had no argument whatever, which applied in the general case.

We will give an inductive proof and then a commentary.

Proof. Base case: For $n = 0$ ($\heartsuit_0$) the left hand side of the formula is 0^2 and the right hand side is $0(0+1)(2 \cdot 0 + 1)/6 = 0$. So they are equal.

Induction step: Let n be given and suppose $\sum_{k=0}^{n} k^2 = \frac{n(n+1)(2n+1)}{6}$ ($\heartsuit_1$). We want to show that $\sum_{k=0}^{n+1} k^2 = \frac{(n+1)(n+2)(2n+3)}{6}$ ($\heartsuit_2$). Isolating the last term of the sum, we have

$$\sum_{k=0}^{n+1} k^2 = \left[\sum_{k=0}^{n} k^2\right] + (n+1)^2 = \frac{n(n+1)(2n+1)}{6} + (n+1)^2$$

by the induction hypothesis ($\heartsuit_3$). Clearly $\frac{n(n+1)(2n+1)}{6} + (n+1)^2 = \frac{(n+1)(n+2)(2n+3)}{6}$ ($\heartsuit_4$) so we have $\sum_{k=0}^{n+1} k^2 = \frac{(n+1)(n+2)(2n+3)}{6}$ ($\heartsuit_5$) as required, completing the induction step.

Thus the identity is true for all n by induction.   □

Commentary: At $\heartsuit_0$ you notice that the induction variable is n, not k. We want to show the equation is true for all n. The letter k is just an index defining the sum. The identity does not "depend on the value of k". In the induction step, the assumption at $\heartsuit_1$, the induction hypothesis, is made for the one particular value of n just given, and only for the purpose of showing the implication. It is not assumed "for all n". That would be a *circular argument*, assuming what you want to prove.

The equation at ($\heartsuit_2$) is not an assumption at all, nor a conclusion. It is an objective to be shown, the consequence of the implication to be established. It is often helpful to the reader to specify a goal, and essential to the writer to know it also, whether it is written into the proof or not. Notice that, yet again, if you leave off the words, the proof is ruined! Since $\heartsuit_2$ is neither assumed nor concluded, you may not use this statement later in the argument.

The core of the argument is at $\heartsuit_3$. The algebraic assertion at $\heartsuit_4$ closing the deal is shockingly short to both those who are weak in algebra, and also to those who glory in it. But the algebraic minutia are not the heart of this proof. The heart is the inductive structure. At $\heartsuit_5$ the induction step is just completed, and, in the notation of Definition 5.5, we have established the implication $P_n \Rightarrow P_{n+1}$ for the given n. At this point, we release the assumption that P_n is true, as well as the conclusion that P_{n+1} is true. Only the implication itself is held to be proved.

Now, for the final paragraph of the proof, since the implication $P_n \Rightarrow P_{n+1}$ was shown for any given n, we actually have $\forall n : P_n \Rightarrow P_{n+1}$, and can match that with the base case to conclude that the identity is true for all $n \geq 0$ by induction.

Another induction example

Here is an unlikely assertion: *The difference between a cubic number and its cube root is always divisible by* 6. At least it seems unlikely until you try a few small cases: $2^3 - 2 = 6$, $3^3 - 3 = 24$, $4^3 - 4 = 60$, $5^3 - 5 = 120$, and skipping ahead, $10^3 - 10 = 990 = 2 \cdot 3^2 \cdot 5 \cdot 11$. Like with the cannonball identity, you cannot check every case one by one, but you can try induction.

Proof. The base case is trivial. Suppose that $n^3 - n = 6k$ $\heartsuit_0$. We need to show that under this condition also $(n + 1)^3 - (n + 1)$ is divisible by 6. Multiplying out we get $(n + 1)^3 - (n + 1) = (n^3 - n) + 3(n^2 + n) = 6k + 3(n^2 + n)$ $\heartsuit_1$ which is clearly divisible by 3. Moreover, $n^2 + n$ must be even since n^2 and n are either both even or both odd. $\qquad\qquad\square$

Commentary: This proof is much harder to read and follow since it leaves much of the inductive structure to the reader to fill in. It is also much harder to write since all the aspects which have been left off must be considered and checked by the writer. The only hint that the proof is an inductive argument is in the sentence preceding the proof. The reader is presumed to be able in fill all in the missing details, and most will.

The most annoying feature of this argument is that the proof refers to a variable n which is not mentioned in the statement of the problem at all! It is often the case, when composing an inductive argument, that the statement to be proved must be recast in terms of a variable on which induction can be done. In this case it would be p_n: *For all $n \in \mathbb{N}$, $n^3 - n$ is divisible by* 6.

The author leaves all the algebra to the reader, $\heartsuit_2$, which is ok for some readers, and at least makes sure that the reader knows that the equation at $\heartsuit_1$ is only assumed. But the reader is never alerted that the induction step is over and the final conclusion is left off as well, so an inexperienced reader has the impression that the proof is unexpectedly over.

But the skeleton is there, and when confronted with an argument like this, the unconvinced reader should at least write out specifically the implication being shown in the body of the argument: $P_n \Rightarrow P_{n+1}$: *If $n^3 - n$ is divisible by 6, then $(n+1)^3 - (n+1)$ is divisible by 6.*

A last bit of commentary. If you want to help out the reader by putting in more algebra details, do *not* do it like this:

$$(n+1)^3 - (n+1) = (n^3 - n) + 3(n^2 + n)$$
$$n^3 + 3n^2 + 3n + 1 - n - 1 = n^3 - n + 3n^2 + 3n$$
$$0 = 0.$$

The problem is not that it is ugly, but that it does not show what is intended. It shows the implication

$$[(n+1)^3 - (n+1) = (n^3 - n) + 3(n^2 + n)] \Rightarrow [0 = 0],$$

which is trivial because the consequence is true. That implication does *not* show that $(n+1)^3 - (n+1) = (n^3 - n) + 3(n^2 + n)$. Something valid in a similar style is

$$(n+1)^3 - (n+1) = n^3 + 3n^2 + 3n + 1 - n - 1$$
$$= (n^3 - n) + 3(n^2 + n)$$

Now that the commentary is over, here is an alternative, non-inductive proof: The expression $n^3 - n = n(n^2 - 1) = (n-1)n(n+1)$ is the product of three consecutive integers. So at least one factor is divisible by 3, and at least one is even. Hence $n^3 - n$ must be divisible by 6.

The non-inductive proof, for the reader, is simpler, more convincing, and gives greater insight. And it proves the result for negative values of n too! For the writer, however, this more direct argument is more difficult to come up with in the first place. For the writer, the inductive argument is easier to make since the implication is easier to show than just the consequence alone. That is why an inductive proof is often the first one to appear for new results, with direct arguments following later, each providing a different bit of insight.

Exercises

1. Rewrite both the statement and the proof of the second example in the style of the first one.
2. Weave the following bit of algebra

$$(n+1)^2 - (n+1) = n^2 + 2n + 1 - 1 = (n^2 - n) + 2n$$

into an induction proof that $n^2 - n$ is always even. Write out carefully the statement to be proved, identify and prove the base case. Include explicitly the implication you establish in the induction step.

3. Suppose the predicate P_n, defined on $n \in \mathbb{N}$, is such that P_{15} is true, P_{14} is false, and $P_n \Rightarrow P_{n+2}$ for all $n \in \mathbb{N}$.

   For each of the following, decide whether it is true, false, or cannot be determined from the given information:

   a) P_0, b) P_4, c) $P_{2^{100}}$, d) P_{55555}.

5.5 Strong induction

Suppose we have a sequence of numbers g_n defined by $g_0 = 1$, $g_1 = 2$, and after that the subsequent values are computed by the formula $g_{n+1} = g_n + 6g_{n-1}$. This type of sequence is called *recursive*. We don't have a closed formula for g_n, but any particular value can be computed by starting from the "bottom" and working up. So the initial values of the sequence are easily computed as $1, 2, 8, 20, 68 \ldots$.

Let's try to prove by induction that $g_n \geq 3^n$.

Proof. For the base case we have to show that $g_0 \geq 3^0$ which is true since $g_0 = 1$.

Now, let $n \geq 0$ be given and assume $g_n \geq 3^n$. We may now compute

$$g_{n+1} = g_n + g_{n-1} \geq 3^n + 6 \cdot 3^{n-1} = 3^{n-1}(3+6) = 3^{n-1} \cdot 3^2 = 3^{n+1}$$

So $g_{n+1} \geq 3^{n+1}$, concluding the induction step.

Therefore the statement is true for all $n \geq 0$ by induction. □

Except that it isn't. $g_4 = 68$, but $3^4 = 81$. In fact, you probably noticed already that the assertion goes bad right away, since $g_1 = 2$ which is less than 3^1. So our predicate is not true for all n. But what is wrong with the proof?

The alert reader may have noticed that the induction step used just a bit more than the induction hypothesis, that $g_n \geq 3^n$. We also used that $g_{n-1} \geq 3^{n-1}$. This assumption is not allowed according to induction as defined in Definition 5.5, but actually is a common practice, sometimes called *strong induction*. Here is how it works. Strong induction, in the induction step, assumes not merely that the nth statement is true, but all the previous statements as well.

In equations:

$$\left[\bigwedge_{n=0}^{\infty} P_n \right] = P_0 \wedge \left[\bigwedge_{n=0}^{\infty} \left[\left(\bigwedge_{k=0}^{n} P_k \right) \Rightarrow P_{n+1} \right] \right] \tag{5.2}$$

or, if you don't want to decode all those indices,

$$P_0, \quad P_0 \Rightarrow P_1, \quad \left. \begin{matrix} P_0 \\ P_1 \end{matrix} \right\} \Rightarrow P_2, \quad \left. \begin{matrix} P_0 \\ P_1 \\ P_2 \end{matrix} \right\} \Rightarrow P_3, \quad \left. \begin{matrix} P_0 \\ P_1 \\ P_2 \\ P_3 \end{matrix} \right\} \Rightarrow P_4, \; \ldots \tag{5.3}$$

The strong induction procedure is justified exactly as the standard version. If all the P_n's are true, then all the implications in Eq. (5.2) are true since the consequences are true. On the other hand, since the base case is true, if any were false, there would be a smallest value of n with P_{n+1} false, but that would mean $P_0 \wedge \cdots \wedge P_n$ is true, violating the implication $\left(\bigwedge_{k=0}^{n} P_k \right) \Rightarrow P_{n+1}$.

So strong induction is valid. But what is wrong with the proof we started this section with? The base case is valid, and the induction step is completely valid too if we are allowed to assume both p_n and p_{n-1}. And strong induction grants you that assumption for all n except $n = 0$, where you have to show that $P_0 \Rightarrow P_1$. That is the problem. The argument for the induction step considers $g_{n+1} = g_n + g_{n-1}$, but there is no g_{n-1} for $n = 0$. We need there a special argument for $n = 0$, the first implication to be shown, and it turns out there isn't one because the result is false. But the rest is perfectly valid reasoning, and can be reused in a slightly different example, in which the induction step for $n = 0$ actually works.

Theorem 5.7. *Let h_n defined by $h_0 = 1$ and $h_1 = 4$ with $h_{n+1} = h_n + 6h_{n-1}$. Then $h_n \geq 3^n$ for all n.*

Proof. The base cases are that $h_0 = 1 \geq 3^0$ and $h_1 = 4 \geq 3^1 = 3$, which are both true.

Now, let $n \geq 1$ be given and assume $h_k \geq 3^k$ for all $0 \leq k \leq n$. We want to show $h_{n+1} \geq 3^{n+1}$. We may now compute

$$h_{n+1} = h_n + h_{n-1} \geq 3^n + 6 \cdot 3^{n-1} = 3^{n-1}(3+6) = 3^{n-1} \cdot 3^2 = 3^{n+1},$$

so $h_{n+1} \geq 3^{n+1}$, concluding the induction step.

Therefore the statement is true for all $n \geq 0$ by (strong) induction.   □

Notice that the $n = 0$ instance of the induction step was folded into the base case, which reads clearer.

Most people using the method of Eq. (5.2) will label their argument an "induction", rather than "strong induction". As we have seen, both methods are equally valid in the sense that any result provable under one method, can be proved, under the other, with at most a variation in how the statement is phrased and indexed. The moral is not to pick one over the other, but to use the one which is convenient to the task.

Exercises

1. Suppose you can show P_n is true for all $n \geq 0$ by strong induction. Show $\bigwedge_{k=0}^{n} P_k$ is true for all $n \geq 0$ by ordinary induction.
2. Let the recursive sequence g_n be defined by $g_0 = 1$, $g_1 = 3$ and satisfying $g_{n+1} = 3g_n - 2g_{n-1}$ for all $n \geq 1$. Show that $g_n \leq 2^{n+1}$ for all $n \geq 0$.
3. Let the recursive sequence h_n be defined by $h_0 = 2$, $h_1 = 3$ and satisfying $h_{n+1} = 3h_n - 2h_{n-1}$ for all $n \geq 1$. Show that $h_n = 2^{n+1} + 1$ for all $n \geq 0$.

5.6 Sets versus logic

If you have been reading this text from the beginning, you certainly must have the feeling that you have done the same thing twice. It's *déjà vu* all over again.

When sets were introduced, we had the notion of well-definition, based on the dichotomy of membership. The key set operations, union, intersection, and complement were introduced. Those operations were connected by identities. Those identities were established using the double inclusion method, and that method was based on the essential $\subseteq$.

But later we considered formal logic, with the dichotomy of TRUE/FALSE and operations AND, OR, and NOT. Virtually the same identities were established with the double implication method, and that method relied on the tricky $\Rightarrow$.

Looked at side by side, it is hard to avoid having the idea that logic and sets are just "the same thing". But, on the other hand, they have a different history, and different notation. Operationally, they seem identical, but conceptionally the subjects are very different. With sets we are considering objects. Linguistically sets and their members are nouns – things we can imagine touching and holding. In set theory we are collecting, distinguishing, arranging, and ordering. With logic, we seem to be in a more abstract, higher realm. In logic we are discussing the truth and falsity of ideas. We feel we are are evaluating, speculating, and judging.

With formal logic, the two subjects, set theory and logic, seem to live in two similar but separate spheres. They seem merely to behave with analogous rules, like baseball and cricket.

With predicate logic, however, we can finally illustrate how, actually, these subjects are connected. Consider predicate logic with predicates defined over the natural numbers. In other words, Boolean functions defined on the set $\mathbb{N}$. Each such Boolean function assigns a value of 0 or 1 to each of the elements of $\mathbb{N}$ and we can imagine the *graph* of that function as a sequence of 0's and 0. The graph for the predicate $n \le 10$ would be

$$1\ 1\ 1\ 1\ 1\ 1\ 1\ 1\ 1\ 1\ 1\ 0\ 0\ 0\ 0\ 0\ 0\ 0\ 0\ 0\ 0\ 0\ 0\ 0\ 0\ 0\ \ldots$$

and isn't that just how we would write the bit vector for the set $\{n \in \mathbb{N} \mid n \le 10\}$. That is the connection. Predicate logic, with the predicates confined to the natural numbers, is equivalent to Boolean functions defined on the natural numbers. That in turn is directly linked by the bit vectors to sets defined on the universe $\mathcal{U} = \mathbb{N}$. Here we really can completely link $\cup$ and $\vee$, $\cap$ and $\wedge$, $(-)^c$ and $\neg$, and mostly importantly, $\subseteq$ and $\Rightarrow$.

Yes, the subjects are different in total, but with the correct restrictions, they are exactly the same.

Exercises

1. Write the set whose bit vector corresponds to the predicate $(n < 4) \wedge (n \neq 2)$ for $n \in \mathbb{D}$.

2. Consider the true implication, "If n is prime and n is even, then $n = 2$". Try to express this using set theory alone.

3. Consider the set $X = \{n \in \mathbb{N} \mid n = 5k + 1; k \in \mathbb{N}\}$. Re-express it as a Boolean function.

5.7   Case study: Decoding the Gray code

The case study of Section 3.9 introduced Gray code $g_n g_{n-1} \cdots g_1 g_0$ of the binary number with bits $b_n b_{n-1} \cdots b_1 b_0$. It is given by the formula

$$g_k = \begin{cases} 0 & \text{if } b_k = b_{k+1} \\ 1 & \text{if } b_k \neq b_{k+1} \end{cases}$$

Note that we naturally take $b_{n+1} = 0$. An alternative, if you want to interpret the 0 and 1 as FALSE and TRUE, is to write $g_k := [b_k \neq b_{k+1}]$. Either formulation allows us to find the bits of the Gray code.

What if we wanted to reverse the process and determine from $g_n g_{n-1} \cdots g_1 g_0$ which ordinal it had, in other words, how many bit vectors are ahead of it in the Gray code order. It is natural to look for a formula for n in terms of its binary bits, b_j, and a bit of trial and error would take you to the following:

Theorem 5.8. $b_k = 0$ *if the number of* 1*'s in the sequence* $g_k, g_{k+1}, g_{k+2}, \ldots$ *is even, and* $b_k = 1$ *otherwise.*

Let's do a quick check. Start with 1776 in binary, 11011110000, and translate to the Gray code using $g_k := [b_k \neq b_{k+1}]$, giving 10110001000. Now, following the reverse procedure of Theorem 5.8, move from left to right recording at each bit the number of 1's encountered so far, which gives 11233334444, and then using 0 and 1 to record even versus odd returns us to 11011110000 as predicted.

The actual proof of the theorem is by induction, but in this case it is a *backwards* induction, starting from the nth bit, and proceeding to the 0th. Of course, we may use in the proof the relation we already have between the bits in the Gray code g_k and the binary bits b_k, namely that $g_k := [b_k \neq b_{k+1}]$.

Proof. Base case: For the nth bit of an n digit binary number, or n digit Gray code, the nth bits correspond, and the claim is true in this case.

Induction step. Let k be given, $n \geq k > 0$, and suppose b_k satisfies the claim. There are four cases for the pair b_k and b_{k-1}.

If $b_k = b_{k-1} = 0$ or $b_k = b_{k-1} = 1$ then it follows that $g_{k-1} = 0$, and the number of 1's in the sequence $g_k, g_{k+1}, g_{k+2}, \ldots$ is exactly the same as in $g_{k-1}, g_k, g_{k+1}, g_{k+2}, \ldots$, and $b_{k-1} = b_k$ correctly records whether that number is even or odd by the induction hypothesis.

On the other hand, if $b_k \neq b_{k-1}$, then $g_{k-1} = 1$, and the number of 1's in the sequence $g_k, g_{k+1}, g_{k+2}, \ldots$ is exactly one fewer than that in g_{k-1}, g_k, $g_{k+1}, g_{k+2}, \ldots$, since b_k correctly measured the evenness of the number of ones in the first sequence by the induction hypothesis and $b_{k-1} \neq b_k$, b_{k-1} correctly measures the parity of the number of 1's in the second. $\square$

There is a fast direct proof which uses some number theory but even that one requires care. Here there are two different binary strings and an index set to be kept in line, together with the inductive variable, so you should not be surprised if you need to read it through more than once to get the complete picture.

5.8   Case study: The 14–15 puzzle

We know that it is often important to distinguish between even and odd numbers. It turns out that permutations also come in two distinct types, also called even and odd. Even or oddness of a string of distinct characters depends on whether the number of pairs of characters in the string that are out of order is even or odd.

Definition 5.9. Given a permutation $\sigma = \sigma_1 \sigma_2 \ldots \sigma_n$ of the ordered set $x_1 < x_2 < \cdots x_n$, we say σ is *even* if the number of pairs (i, j) with $i < j$ but $\sigma_i > \sigma_j$ is even. Otherwise σ is said to be *odd*. ♠

So for example $abcde$ is even because all pairs are in order. For $edcba$ all pairs are out of order, but there are $\binom{5}{2} = 10$ of them, so $edcba$ is also even. Reversing four characters $dcba$ is also even since the $\binom{4}{2} = 6$ is even, but $fedcba$ is odd because $\binom{6}{2}$ is 15. The number of out of order pairs is called the *inversion number* of the permutation. In general it can be computed by counting, for each character, how many characters to the right of it precede it in the ordering on the set. So for $bcedfga$ we have $1 + 1 + 2 + 1 + 1 + 1 + 0 = 7$ which tells us it is an odd permutation. You may recognize the numbers in the sum since we computed exactly the same values to determine that $bcedfga$ had exactly $1 \cdot 6! + 1 \cdot 5! + 2 \cdot 4! + 1 \cdot 3! + 1 \cdot 2! + 1 \cdot 1! + 0 \cdot 0!$ permutations ahead of it in lexicographic order.

There is a key observation about the evenness of permutations, which we may prove by induction, which involves transposing pairs in a string. *Transposing* two characters in permutations exchanges their positions. So transposing 3 and 5 in 6234157 gives the permutation 6254137. Of course, if you transpose a pair twice, you return to the original.

Theorem 5.10. *Given any permutation, transposing any two characters transforms the permutation from even to odd, or vice versa.*

Proof. Suppose we have a permutation σ of n characters. We will do induction on how far apart the characters being transposed are.

Base case. If the characters transposed are next to one another, then transposing them puts them in order if they were out of order, and puts them out of order if they were in order. Other than that, all other pairs have the same relative position and have the same order type they had before. Thus the total number of misordered pairs is changed by exactly one, and the inversion number is changed from even to odd, or the reverse.

Induction step. Let k be given and suppose the result is true for all transpositions separated by k characters. Suppose the pair σ_i and σ_j are separated by $k + 1$ characters. Then the permutation σ can be written $\sigma = X\sigma_i Y\sigma_m\sigma_j Z$ with the string Y having k characters. Make the following three transpositions:

$$X\sigma_i Y\sigma_m\sigma_j z \rightarrow X\sigma_m Y\sigma_i\sigma_j z \rightarrow X\sigma_m Y\sigma_j\sigma_i z \rightarrow X\sigma_j Y\sigma_m\sigma_i Z.$$

Each transposition switches the evenness/oddness of the inversion number, either due to the base case or the induction hypothesis. So, having been switched three times, if the permutation was odd it is now even, and vice versa.

Thus the result holds for all transpositions of σ by induction. □

It is not hard to see, and there is an easy proof by induction, that every permutation can be obtained by a sequence of transpositions. A similar argument shows that it is possible to do the job in at most $n - 1$ transpositions. Try it. Since every permutation can be made just using transpositions, and since the start is the even permutation of all the characters in the correct order, we have an alternate way of viewing the evenness and oddness of a permutation.

Theorem 5.11. *Every permutation is even or odd depending on whether the number of transpositions needed to express it is even or odd.*

A more dramatic way of saying this is to notice that every permutation can either be created using an even number of transpositions, or an odd number of transpositions, *but never both*. Most people find this result quite strange, even if they can follow the proof.

In the 14–15 puzzle, Sam Loyd, the puzzle master of the 1800's made curious device with 15 sliding numbered squares and one blank, held loosely in a 4×4 frame. Because of the blank, it was possible to slide the pieces about and achieve many permutations of the original configuration of numbers. Loyd offered a large cash prize to anyone who could take the puzzle from the initial state in which he sold it, with the numbered squares in order except that the final squares 14 and 15 were transposed, to a completely ordered state. See Fig. 5.1.

Loyd knew that he was asking the solver to effect an odd permutation of the squares. He also knew that every move the solver made was to transpose the blank with some actual tile. Also, he knew that if the blank was to return to its original position, no matter what happened, the number of times it moved up had to match exactly the number of times it moved down, and the number of times it moved to the left had to match the number of times it moved back to the right. So the only way to return the blank to its original position was through

FIGURE 5.1 Sam Loyd's 14–15 puzzle. Initial position on the left.

an even number of transpositions. Loyd knew not only that it was impossible for anybody to ever be able to bring the squares into order and collect the prize money, but also that very few people would be able to understand why it was impossible and why they were failing over and over and over.

It is easy to find inexpensive plastic versions of the sliding squares mechanism, but they are always sold with the squares in the proper order, with the idea that children can amuse themselves by disordering and reordering the squares. If you want to use one of these toys but achieve the same feeling of frustration that Loyd's contestants did, then try to return the blank to the original position with the squares in reverse order. That is an odd permutation of the squares since $\binom{15}{2} = 105$.

5.9   Case study: Towers of Hanoi

In the previous case study on the Towers of Hanoi we used the multiplicative principle to conclude that, for 7 disks, there are 3^7 legal states. If we label the three posts, left to right, as 0 for hell, 1 for earth, and 2 for heaven, then the states correspond to the 7 digit numbers in base three, see Fig. 5.2. Let's consider one of the key problems: is it possible to move all the disks from hell, state 0000000, to heaven, state 2222222, and if so, in how many moves?

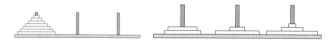

FIGURE 5.2   The Towers of Hanoi with 7 disks in states 0000000 and 2100212.

Counterintuitively, one way to simplify the problem is to expand the setup to include versions with the number disks not restricted to seven. In other words, to consider the same problem, still with three posts, but with d disks, $d \in \mathbb{N}$. While this expands the range of possibilities, it allows for an inductive approach.

Now it takes just a few seconds to determine that you can solve the problem with 1 disk in 1 move, and the problem with 2 disks in 3 moves. Since there are three posts, you might expect that three disks will require 9 moves, and indeed it can be done in 9 moves, but you can also do it in only 7. Not much of a saving – or is it?

✠

If you have paused to come up with a formula on your own, you probably got the formula $2^d - 1$, and that is what we will prove.

Theorem 5.12. *In the d-disk Towers of Hanoi problem, the disks may be moved from state 0 to state $3^d - 1$ in at most $2^d - 1$ moves.*

Note that, since the states are encoded with ternary numbers, state $3^d - 1$ has all d disks on post 2, and state 0 has all d disks on post 0. As expected, the proof is by induction on d, and the base case will be $d = 1$. (Even though the formula sort of works for $d = 0$ too.)

Proof. For $d = 1$, the disk is unrestricted and can be moved from hell to heaven in one move.

Let d be given and suppose the theorem is true for d disks. We want to show that a tower of $d + 1$ disks can be moved in at most $2^{d+1} - 1$ moves. Temporarily swapping the labels on posts 1 and 2, the induction hypothesis says that we can move the top d disks from hell to earth in at most $2^d - 1$ moves. Now the bottom disk is free to be moved to heaven, where it can stay for the remainder of the procedure without violating any rules. Then, relabeling posts 1 and 0, the induction hypothesis allows us to move the d disks on earth to heaven in at most another $2^d - 1$ moves. See Fig. 5.3.

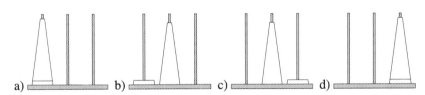

FIGURE 5.3 $a \to b$: $2^d - 1$ moves. $b \to c$: 1 move. $c \to d$: $2^d - 1$ moves.

Altogether we have taken at most $(2^d - 1) + 1 + (2^d - 1) = 2 \cdot 2^d - 1 = 2^{d+1} - 1$, as required.

So the theorem is true for all $d \geq 1$. $\qquad\qquad\square$

Virtually the same proof, and the same figure, proves the following companion observation. This is a tribute to the power of the inductive method, which, with a little care, can leverage many results from one very small insight.

Theorem 5.13. *In the d-disk Towers of Hanoi problem, the disks cannot be moved from state 0 to state $3^d - 1$ in fewer than $2^d - 1$ moves.*

Try to write out the proof. You just need to modify the words in the first proof, leaving all the equations and variables right where they are. These two theorems work together to tell us that the *optimal solution* takes exactly $2^d - 1$ moves.

Can you show it is the *unique* optimal solution?

So, what if we start in some other legal position? Here is an informal induction that we can again stack everything on post 0 in at most $2^d - 1$ moves. The base case is trivial. Assume it works for d disks, and suppose we have a legal position with $d + 1$ disks. If the largest disk is on post 2, leave it there, and we can move the remaining disks on top of it in at most $2^d - 1$ moves by the induction hypothesis, and $2^d - 1 < 2^{d+1} - 1$. If the largest disk is not on post 2 then, by the induction hypothesis, we can move all the other disks legally on the other post, not 2 and not the one containing the largest disk. Then move the largest disk onto post 2, and use the induction hypothesis again to pile the smaller disks on top of it, altogether a total of, at most, $(2^d - 1) + 1 + (2^d - 1) = 2^{d+1} - 1$ moves.

This is essentially the same argument again! Now, what if the disks are stacked illegally, say an evil Genie randomly dropped the disks on the posts regardless of the rules, but we still want to move them to heaven using only legal moves. Can it be done? How quickly?

Here is a completely different question. For $d = 7$, the fraction of legal states used in the optimal solution is $(2/3)^7$, less than 6% of the states. How do we recognize the states in the optimal solution? Or, given a state, what is the best move to make? How many problems can be solved by just tweaking the original inductive proof? There are so many interesting questions one can ask.

5.10 Case study: The Fibonacci numbers

The recursive sequences in Section 5.5 might have reminded you of the Fibonacci sequence. The sequence is thought of as being generated by the monthly population of mathematical rabbits whose numbers are strictly governed by the following simple rule: *Each month, every pair of rabbits which is at least one month old mates and generates a pair of leverets (baby rabbits) one month later.*

Since the rabbits have no predators, in fact are presumed to be immortal and require no food, the population grows and grows. At month 0 the population is zero, but at month 1 the first pair of leverets arrive (by mail order) and the process begins.

By the rule, the number of pairs of leverets produced at the end of month $n + 1$ is equal to the number of pairs of rabbits mating during month n, which is the same as the number of pairs existing in month $n - 1$. Also, the number of pairs of adult rabbits in month $n + 1$ is equal to the number of pairs of rabbits of any kind existing in month n. Adding these gives the total number of pairs alive in month $n + 1$. Letting f_n be the number of pairs of rabbits in month n, the situation is summarized in the following recursive definition.

Definition 5.14. The *Fibonacci sequence* is obtained recursively from the equations

$$f_0 = 0, \quad f_1 = 1, \quad f_{n+1} = f_n + f_{n-1}. \quad \spadesuit$$

The sequence starts innocently enough, and at the end of six months the number of pairs is still in the single digits, but after only two years there are almost fifty thousand pairs:

$$0, 1, 1, 2, 3, 5, 8, 13, 21, 34, 55, 89, 144, 233, 377,$$
$$610, 987, 1597, 2584, 4181, 6765, 10946, 17711, 28657, 46368\ldots$$

The growth looks exponential, and that is one way of analyzing the behavior. In any case, the growth depends on the fraction of the population which is at least one month old, f_{n-1}/f_n, which, after the first few months have passed, seems to be about $2/3$, and is at least $1/2$.

Theorem 5.15. *For $n \geq 2$, $f_{n-1}/f_n \geq 1/2$.*

Of course we prove this by induction.

Proof. Base case: If $n = 2$ then $f_1/f_2 = 1/2$, which is spot on, and for $n = 3$, then $f_2/f_3 = 2/3 \geq 1/2$.

Induction step: Let $n \geq 3$ be given and assume $f_{n-1}/f_n \geq 1/2$ and $f_{n-2}/f_{n-1} \geq 1/2$. Rearranging the inequalities gives $f_n \leq 2f_{n-1}$ and $f_{n-1} \leq 2f_{n-2}$, and adding these gives $f_n + f_{n-1} \leq 2f_{n-1} + 2f_{n-2}$, or $f_{n+1} \leq 2f_n$ using the recursive formula. Thus $f_n/f_{n+1} \geq 1/2$, as required.

So the theorem is proved by induction. □

You should check that the inductive argument is correct, that the base cases are correct and fit together with the strong induction step.

By the theorem, for months beyond 2, the next month's rabbits will be at least the current month's rabbits, plus the number of leverets born to at least half the population, altogether at least $1 + 1/2 = 3/2$ of the current population. This gives an exponential lower bound on the Fibonacci sequence:

$$f_{n+2} \geq f_2(3/2)^n.$$

Perhaps one of the most surprising results about the Fibonacci numbers is the equation

$$f_n = \frac{(1 + \sqrt{5})^n - (1 - \sqrt{5})^n}{2^n \sqrt{5}} \tag{5.4}$$

The proof by induction just follows the model above, but it takes persistence and there is some tricky algebra. Even if you get through the proof, the result still looks quite mad. Only the cases $n = 0$ and $n = 1$ seem easy to verify. Still, it is a good exercise to verify that

$$f_5 = 5 = \frac{(1 + \sqrt{5})^5 - (1 - \sqrt{5})^5}{2^5 \sqrt{5}}$$

using the Binomial Theorem. And, if you really like algebra, you can try to find a direct non-inductive proof of Eq. (5.4) using only the Binomial Theorem.

5.11   Summary exercises

You should have learned about:

- Predicates and Boolean functions on general sets.
- Quantification of Predicates, and Evaluations of Boolean functions.
- Universal and Existential Quantification.
- What are the requirements of an induction proof.
- Why an induction proof is valid.
- The requirements of a strong induction proof.
- Why a strong induction proof is valid.
- How to write a simple induction proof.
- How to follow an argument that is based on induction.

1. Let $P(n)$ be the predicate $\sum_{k=0}^{n} k^2 = \frac{n(n+1)(2n+1)}{6}$. Show each of the following:

    a) $\exists n \in \mathbb{N}; P(n)$,   b) $P(10) \Rightarrow P(11)$,   c) $P(22) \Rightarrow P(11)$.

    [Hint: You can show all three by checking the validity of $P(n)$ for a single value $n \in \mathbb{N}$, and it is not $n = 0$.]

2. For $n \in \mathbb{N}$, let A_n be a set and let P_n be the predicate "$|A_n| < \binom{3n+1}{n+1}$".
    Write out each of the following:

    a) P_0 b) P_9 c) $P_n \wedge P_{n+2}$ d) $P_n \Rightarrow P_{n+1}$.

3. Suppose you wanted to prove that $\sum_{k=1}^{n}(2k-1) = n^2$.
    Identify the inductive variable.
    What is the base case?
    What implication would you have to show for the induction step?

4. Suppose p_n and q_n are predicates defined for all $n \geq 0$, and you wanted to show $\forall n \in \mathbb{N}; (p_{n^2+1} \vee q_{2n})$ by induction.
    What would you have to show for the base case?
    What would you have to show for the induction step?
    What would be the induction hypothesis?

5. Let $P(n)$ be a predicate defined on the natural numbers.
    Suppose $\forall n \in \mathbb{N}; [P(n) \Rightarrow P(n+1)]$ and suppose $P(10)$ is false, $P(100)$ is false and $P(1000)$ is true. Label each of the following T if it must be true, F if it must be false, and X if it cannot be certainly concluded. Give a word of explanation:

    $P(15), \quad P(150), \quad P(1500), \quad P(1776)$.

6. $(p_n \Rightarrow p_{n+2}) \wedge (p_n \Rightarrow p_{n+3})$. What can we conclude by induction? Why?

7. Suppose p_0 is true and that for all $n \in \mathbb{N}$, we have $(p_n \Rightarrow p_{n+5}) \wedge (p_n \Rightarrow p_{n+11})$. What can we conclude by induction? Why?

8. Suppose p_0 and $\forall n \geq 1; [(p_n \Rightarrow p_{n+1}) \vee (p_{n-1} \Rightarrow p_{n+1})]$. Which p_n can we conclude are true? Why?

9. Suppose you have a predicate p_n for $n \in \mathbb{N}$. Suppose that p_0 is true and suppose that, for all $n \geq 0$, we have $p_n \Rightarrow (p_{n+1} \vee p_{n+2} \vee p_{n+3})$.
    Prove by induction that $p_n \vee p_{n+1} \vee p_{n+2}$ is true for all n.

10. Let $e_0 = 1$ and $e_1 = 3$, and define $e_{n+1} = 5e_n - 6e_{n-1}$ for $n \geq 1$. Prove $e_n = 3^n$ for all $n \geq 0$.

11. Let $e_0 = 0$ and $e_1 = 1$, and define $e_{n+1} = 5e_n - 6e_{n-1}$ for $n \geq 1$. Prove $e_n = 3^n - 2^n$ for all $n \geq 0$.

12. Let the recursive sequence g_n be defined by $g_0 = 2$, $g_1 = 5$ and satisfying $g_{n+1} = g_n + 6g_{n-1}$ for all $n \geq 1$. Show that for all $n \geq 0$, the formula $g_n = [3^{n+2} + (-2)^n]/5$ is valid.

Chapter 6

Set structures

6.1 Relations

With mathematical sets to describe our objects of interest, and predicate logic to communicate what is true about them, we can now be said to have reached a critical juncture in our study of discrete mathematics, "this is not the end, this is not even the beginning of the end, this is just perhaps the end of the beginning". From here one can see, stretching out in many directions and with much intertwining, the many and various branches of discrete mathematics. But wherever you travel, logic and sets will be the underpinning.

This chapter is the introduction to a huge area of set structures, in which sets and their elements are not just to be thought of in isolation, but in concert with other sets. There are many forms of set structures; a vast array of data structures, algebraic structures, combinatorial structures, geometrical and topological structures, only a few of which we will touch upon. We have already seen several set structures, such as strings and bit vectors, both data structures. Those are key examples because they illustrate the two most common ways of building set structures – via subsets and via the Cartesian product.

In this chapter we will concentrate on *relations*, arguably the most basic example. 'Set structure' is an idea, like discreteness, but in mathematics "relation" is not the vague term which floats about in natural language, it has a set mathematical definition, which you must learn.

Definition 6.1. A (binary) *relation*, R, between sets A and B is a subset $R \subseteq A \times B$. If $(a, b) \in R$ then we say *a is related to b by R*. ♠

That seems very innocent, but it is surprisingly powerful. If you are programming a chess game, then you have many sets of interest, certainly the pieces P and the set of squares S. The most basic relation would be the *board relation* $B \subset P \times S$ with the elements of B being those pairs (p, s) for which piece p is currently on square s.

Here is a more mathematical example:

$$L = \{(n, m) \in \mathbb{Z} \times \mathbb{Z} \mid n < m\}$$

So 3 is related to 5 but 5 is not related to 3 by L. You should recognize this relation as "less than". A relation which becomes established over time is usually not notated as a set, but acquires a special symbol, in this case $<$. One also commonly writes $2L18$ or $2 < 18$, with the symbol *infix*, instead of writing $(2, 18) \in L$ or the, admittedly ugly, $(2, 18) \in <$. Many of the common mathematical symbols you know, $\in$, $\subseteq$, $\neq$, etc. are examples of relations, and are usually written in infix notation.

Another common notation, which is particularly valuable for small sets, is the *relation diagram:* You place a labeled point for each element $a \in A$, and for each element $b \in B$, and draw an arrow between the points associated with a and b if a is related to b by R, that is if $(a, b) \in R$, that is, if $a R b$.

Example 6.2. In Fig. 6.1 we have a set of four common household ingredients and a set of four recipes in which they are *used*, and the diagram defines a relation U. So we have that (eggs, treacle) $\notin U$, but milk$U$ice cream. For small sets it is not hard to make a diagram which is far easier to understand than listing the elements in the relation set. ◇

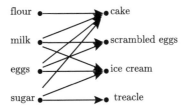

FIGURE 6.1 A relation graph.

The same relational information may be conveyed by $(a, b) \in R$ (set theoretic), a is related to b by R (textural), $a R b$ (infix), and the diagrammatic notation may even be used in a paragraph, like $\overset{a}{\bullet} \longrightarrow \overset{b}{\bullet}$ (diagrammatic).

Relations are used so widely they come in many special forms. In the rest of this chapter we consider various special types of relations.

Exercises

1. Let $R \subseteq \mathbb{A} \times \mathbb{D}$ be the relation between the set $\mathbb{A}$ of letters and the set $\mathbb{D}$ of digits by setting $a R d$ if the letter a occurs in the spelling of the English word for the digit d.
   Draw the relation diagram for R.

2. Let $\heartsuit$ be the relation between $\mathcal{P}_3(\{0, 1, 2, 3, 4\})$ and $\mathcal{P}_2(\{2, 4, 6, 8\})$ by setting $A \heartsuit B$ if two elements of A sum to an element of B. Draw the relation diagram of $\heartsuit$.

3. Consider all relations between the set $X = \{2, 4, 8\}$ and itself. How many relations do not have any element related to itself?

6.2   Functional relations

Definition 6.3. A relation $F \subseteq D \times T$ is said to be *functional* if $D \neq \emptyset$ and for each element $d \in D$ there exists exactly one element $t \in T$ to which it is related. The set D is called the *domain* set, and the set T is called the *target* set.   ♠

The relation diagrams of functional relations have the property that there is exactly one outgoing arrow from each element of the domain set. There is no condition on the incoming arrows to the target set. In Fig. 6.2 the third relation is not functional because domain element d_3 is unrelated, and the fourth relation is non-functional because the domain element d_2 is trebly related.

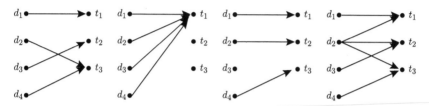

FIGURE 6.2   Four relations, two functional and two non-functional.

If the relation $F \subseteq D \times T$ is functional, it is often called simply a *function*, and the notation $F : D \to T$ is used to specify the domain and target sets. Also, the functional restriction, focusing so much attention on the domain side, gives rise to two other notations for the related elements. If $(d, t) \in F$, then we write $F(d) = t$ in the *operator notation* and $F_d = t$ in the *subscript notation*.

The operator notation is by far the most common notation for functional relations (functions) as found in high school algebra, pre-calculus, and calculus. Please note, functions in high school and calculus are almost always given by an algebraic expression, which is hugely important there, while the specific domain and target sets are usually of secondary importance. In discrete mathematics, most of our functional relations are not given by formulas or algebraic expressions, and the domain and target sets are often the main objects of interest. In particular, the functional definition above, and the definitions of one-to-one and onto which you find below, cannot be glossed over in discrete mathematics. The successful student must attend to their precise definitions.

The notions of *one-to-one* and *onto* arise from applying the two aspects of the functional relation condition to the target set, rather than the domain.

Definition 6.4. A functional relation $F \subseteq D \times T$ is said to be *onto* if for each element $t \in T$ there exists **at least** one element $d \in D$ to which it is related.

A functional relation $F \subseteq D \times T$ is said to be *one-to-one* if for each element $t \in T$ there exists **at most** one element $d \in D$ to which it is related.

A functional relation $F \subseteq D \times T$ is said to be *one-to-one and onto* if for each element $t \in T$ there exists **exactly** one element $d \in D$ to which it is related.

♠

Note that these notions are only defined for functional relations. In Fig. 6.3 we see the diagrammatic features of the one-to-one and onto functional relations. The onto relation has at least one incoming arrow at each target node, and the arrows of the one-to-one diagram are all completely separated from one another.

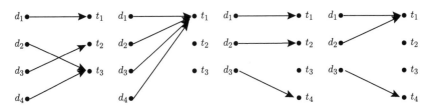

FIGURE 6.3   Four functional relations, one onto, and one one-to-one.

Example 6.5. For the set of pieces in play in the game of chess, the relation between those pieces and the 8^2 positions they can occupy on the board is functional, because each piece occupies exactly one square on the board. The relation is also one to one because each square on the board can have at most one piece. The relation is not onto, because, no matter the configuration, there are sixty-four positions and at most thirty-two pieces in play. So there will always be target positions without pieces.

The relation between the pieces in play and the colors black and white, indicating to which side the piece belongs, is also functional. Each piece is assigned exactly one of the two colors. The relation is also onto since the two kings, one of each color, are never removed from the board, so both target colors are represented. The only way for the relation to be one-to-one is in a stalemate configuration with only the two kings, one of each color.

If the positional relation is expanded to the whole set of 32 chess pieces, the positional relation will often fail to be functional. As soon as a piece is captured, that piece is removed from the board and so has no position related to it. (If a competing chess analyst insists on a functional relationship, he may define a special 65th position, off the board, to act as the "graveyard" for the captured pieces.) $\diamond$

The obvious cardinality conditions mentioned in the example occur quite often, and are stated here as a theorem.

Theorem 6.6. *Let D and T be finite sets. If $|D| > |T|$, then there can be no one-to-one function $f : D \to T$. If $|D| < |T|$, then there can be no onto function $g : D \to T$.*

Exercises

1. Define a relation on $\mathbb{Z} \times \mathbb{Z}$ which has no element related to itself, and is not functional.

2. Define an onto function $g : \mathcal{P}(\mathbb{N}) \to \mathbb{N}$.
   Is your function one-to-one?

3. Define a function $z : \mathbb{N} \to \mathbb{N} \times \mathbb{N}$ by setting $z(n) = (a, b)$ with a the number obtained by deleting the digits in the even positions of the decimal representation of n, and b the number obtained by deleting the digits in the odd positions of the decimal representation of n. So $z(5280) = (58, 20)$.
   Is the function z onto? Is the function z one-to-one? (Hey! – Is it a function?)

6.3 Counting functions on finite sets

In Example 6.5 we saw that whether a relation is functional or not depends not just on the nature of the relationship between the elements, but exactly which sets are considered to be related. In this section we count the number of functions of each type that can exist between a finite domain, and a finite target.

All sets in this section are finite.

Theorem 6.7. *Let D and T be finite sets. The number of functions with domain D and target T is $|T|^{|D|}$.*

Of those, if $|D| \leq |T|$, there will be $\dfrac{|T|!}{(|T| - |D|)!}$ one-to-one functions.

If $|D| \geq |T|$, then $\displaystyle\sum_{k=0}^{|T|} (-1)^k \binom{|T|}{k} (|T| - k)^{|D|}$ functions will be onto.

If $|D| = |T|$, then $|D|! = |T|!$ will be both one-to-one and onto.

The results for general functions and the one-to-one functions are an easy consequence of the multiplicative principle, choosing target elements for each domain element, using straight independence for general functions, and weak independence for one-to-one.

For a function that is both one-to-one and onto, we must have that $|D| = |T|$, and $|T|!/(|T| - |D|)! = |T|!$. In this case we also have the following special situation, whose proof is left as an exercise.

Theorem 6.8. *For finite sets with $|D| = |T|$. Every one-to-one function is also onto, and every onto function is also one-to-one.*

It is a common error to misread this; the theorem does not say that, if $|D| = |T|$, then all functions are one-to-one and onto. The only case where all functions are one-to-one is $|D| = 1$, and the only case where all functions are onto is $|T| = 1$.

Counting the number of onto functions has been left for last, because that is the hardest case. Let us compute the number of non-onto functions by inclusion/exclusion. Let $t \in T$ and define $\mathcal{F}_t$ to be set of all functions $D \to T$ for which the target element t is unrelated. The set we want to count, the set of non-onto functions, is the union $\bigcup_{t \in T} \mathcal{F}_t$ and inclusion/exclusion gives, taking

$T = \{1, 2, 3, \ldots, n\}$ for simplicity.

$$\left| \bigcup_{t \in T} \mathcal{F}_t \right| = |\mathcal{F}_1| + |\mathcal{F}_2| + \cdots + |\mathcal{F}_n|$$
$$- |\mathcal{F}_1 \cap \mathcal{F}_2| - |\mathcal{F}_1 \cap \mathcal{F}_3| - \cdots - |\mathcal{F}_{n-1} \cap \mathcal{F}_n|$$
$$+ |\mathcal{F}_1 \cap \mathcal{F}_2 \cap \mathcal{F}_3| + |\mathcal{F}_1 \cap \mathcal{F}_2 \cap \mathcal{F}_4| - \cdots + |\mathcal{F}_{n-2} \cap \mathcal{F}_{n-1} \cap \mathcal{F}_n|$$
$$\vdots$$
$$+ (-1)^{n-1} |\mathcal{F}_1 \cap \mathcal{F}_2 \cap \cdots \cap \mathcal{F}_n|$$

The k-fold intersection terms, like $|\mathcal{F}_1 \cap \mathcal{F}_2 \cap \ldots \cap \mathcal{F}_k|$, count the functions which omit k elements from the target. Each has cardinality $(|T| - k)^{|D|}$, and there are $\binom{n}{k}$ such terms, so for the k-fold terms we are altogether including or excluding $(|T| - k)^{|D|} \binom{n}{k}$. This justifies the last formula of Theorem 6.7.

Example 6.9. Consider the two sets $A = \{1, 2, 3, 4\}$ and $B = \{a, b, c, d, e\}$. There are 5^4 functions with domain A and target B, and 4^5 functions with domain B and target A.

There are $5!$ one-to-one functions with domain A and target B, and no onto functions.

There are

$$1 \cdot 4^5 - 4 \cdot 3^5 + 6 \cdot 2^5 - 4 \cdot 1^5 + 1 \cdot 0^5$$

onto functions with domain B and target A, and no one-to-one functions. ◇

Exercises

1. Let $A = \{1, 2, 3, 4, 5\}$ and $B = \{a, b, c, d\}$.
   How many functions are there from $\mathcal{P}_2(A)$ to $\mathcal{P}_2(B)$?
   How many are one-to-one?
   How many are onto?
   How many are one-to-one and onto?
2. How many functions are there with domain and target both $\mathcal{P}(\mathcal{P}(\mathcal{P}(\emptyset)))$?
   How many are one-to-one?
   How many are onto?
   How many are one-to-one and onto?
3. Let $A = \{a, b, c, d, e\}$, and $B = \{1, 10, 100, 1000\}$. How many functions are there from $B \times B$ to $A \times A$ which are one-to-one but not onto?

6.4 Working with functional relations

The functional condition is often split into two aspects,

$$\forall d \in D; \exists t \in T; (d, t) \in f, \tag{6.1}$$

$$\forall d \in D;\ \big[[((d,t) \in f) \wedge ((d,t') \in f)] \Longrightarrow (t=t')\big], \tag{6.2}$$

the *existence* of the related target element, Eq. (6.1) and the *uniqueness* of the related target element, Eq. (6.2).

These conditions, with domain and target reversed, are the conditions for one-to-one and onto. A function $f : D \to T$ is one-to-one if

$$\forall t \in T;\ \big[[((d,t) \in f) \wedge ((d',t) \in f)] \Longrightarrow (d=d')\big], \tag{6.3}$$

and the function f is onto if

$$\forall t \in T; \exists d \in D;\ (d,t) \in f. \tag{6.4}$$

So, to properly understand how to work with these definitions, it is enough to understand how the last two work.

To show a function is onto, that is to show that Condition (6.3) is true: Let the target element be given, show it is related to some domain element.

To show a function is not onto: Exhibit any target element not related to a domain element.

Example 6.10. Let $f : \mathcal{P}(\mathbb{D}) \to \mathbb{D}$ be defined by relating each non-empty set $A \subseteq \mathbb{D}$ to be the smallest element in A. So $f(\{3,4,5\}) = 3$. In order for f to be functional, a value must be assigned to $\emptyset$, so define $f(\emptyset) = 0$.

To show f is onto, let $n \in \mathbb{D}$ be given, then $f(\{n\}) = n$. Note that we only have to produce one element of the domain whose function value is the general target digit n. For the onto question we don't to consider whether there are others. Of course, there may be many other quite different but valid arguments. ◇

Example 6.11. Let $g : \mathbb{D} \to \mathcal{P}(\mathbb{D})$ be defined by mapping each digit d to the set of even digits if d is even and to the set of odd digits if d is odd. Convince yourself that g is a function. It is neither one-to-one nor onto. Since the cardinality of the target set is larger than the cardinality of the domain, g cannot be onto. On the other hand, g is not one-to-one, because $g(1) = g(3) = g(5) = g(7) = g(9) = \{1,3,5,7,9\}$. It is equally valid to note only that $g(2) = g(4) = \{0,2,4,6,8\}$. ◇

To show a function is one-to-one: Let two domain elements be given with the same target, $f(d) = f(d')$, and show those two domain elements must coincide, $d = d'$.

To show a function is not one-to-one: Exhibit any pair of *distinct* domain elements which are related to the same target element.

Neither of the previous examples were one-to-one. For f, we have the violating pair $f(\{1\}) = f(\{1,2,3\}) = 1$. For g, we can take $g(1) = g(3) = \{1,3,5,7,9\}$.

Example 6.12. Let $h : \mathbb{D} \to \mathcal{P}(\mathbb{D})$ be defined by $h(d) = \{0, 1, \ldots, d\}$. So $h(0) = \{0\}$ and $h(9) = \mathbb{D}$. The function h is not onto since no digit is related to $\{1\}$. Indeed, $0 \in h(d)$ for all digits d.

To show h is one-to-one, let $h(d) = h(d')$ be given. So $\{0, 1, 2, \ldots, d\} = \{0, 1, 2, \ldots, d'\}$ and, since the sets have the same largest elements, $d = d'$, as required. $\diamond$

Exercises

1. Let $D = \{0, 1, 2, 3, 4, 5, 6, 7, 8, 9\}$. Define a function

$$h : \mathcal{P}_8(D) \to \mathcal{P}_2(D)$$

   by setting $h(X)$ to be the subset consisting of the largest and smallest elements of X. So $h(\{0, 1, 2, 3, 4, 5, 6, 7\}) = \{0, 7\}$.
   Show that h is not onto.
   Define any onto function, g, from $\mathcal{P}_8(D) \to \mathcal{P}_2(D)$.
2. For the function h of the previous exercise, show that h is not one-to-one.
3. Let $D = \{0, 1, 2, 3, 4, 5, 6, 7, 8, 9\}$. Define a function $f : \mathcal{P}(D) \to \mathbb{N}$ by setting $f(\emptyset) = 0$ and $f(A)$ to be the number of different numbers which can be expressed in some base b, with $1 < b \le 10$ using the digits in A, each at most twice.
   So, for $f(\{1, 2\})$, we would consider numbers like 2 or 22 or 1212 all in base 10, or 5 in base 3, $5 = 12_3$, or 10 in base 8, $10 = 12_8$, $56 = 211_5$, etc.
   Show that f is not onto.

6.5 Functions on infinite sets[†]

Working with functions on infinite sets is in many ways similar to the finite case.

Example 6.13. Let the function $f : \mathbb{Z} \to \mathbb{N}$ be defined by

$$f(n) = \begin{cases} 2^n & n > 0 \\ 3^{-n} & n \le 0 \end{cases}.$$

Show that f is one-to-one.

Suppose $f(n) = f(n')$. If $m = f(n)$ is even then it is a power of 2, thus $2^n = 2^{n'}$, so $2^{n-n'} = 1$, and $n - n' = 0$, so $n = n'$, as required. On the other hand, if $f(n)$ is odd, then it can only be a power of 3, and the same algebra starting with $m = 3^{-n} = 3^{-n'}$, gives the result in this case as well. $\diamond$

Example 6.14. Let the function $f : \mathbb{N} \to \mathbb{Z}$ be defined by

$$g(n) = \begin{cases} n/2 & \text{if } n \text{ is even} \\ -(n+1)/2 & \text{if } n \text{ is odd} \end{cases}.$$

Show that f is onto.

Let $m \in \mathbb{Z}$ be given. If $m \geq 0$, then $2m$ is an even natural number and $g(2m) = m$. If $m < 0$, then $-(2m - 1)$ is an odd natural number, and $g(-(2m - 1)) = -((-2m - 1) + 1)/2 = -(-2m)/2 = m$. So m is related to an element of the domain in either case and g is onto. $\diamond$

Both of these examples are counterintuitive, in the sense that, for the first, we have a one-to-one function from a set to a smaller subset, and in the second, an onto function from a subset to a larger set. This situation could not happen for finite sets, and is sometimes called a *Hilbert's Hotel* phenomenon.

You can think of the first example as a way of assigning an infinite number of guests, labeled by the integers, to the infinite number of beds, labeled just by the natural numbers, in the Hilbert Hotel. Each guest is assigned exactly one bed, so the relation is functional. Moreover, at least for the function f, no bed is assigned to two people – a one-to-one function. It sounds impossible, since there are infinitely many more guests than beds. But the one-to-one function f manages the task and even has an infinite number of beds left over! The sequence of beds labeled from 100 to 200 is almost completely unused. The only power of 2 in that range is $2^7 = 128$. So the 7th positive guest has that whole wing to himself since the powers of 3 skip those beds completely, going from bed $3^4 = 81$ to bed $3^5 = 243$.

The second example seems just as strange in the hotel interpretation. Here the guests are labeled by the natural numbers, and they show up at an even bigger Hilbert Hotel with beds labeled from the set of integers. But the function g is onto, so somehow, under this assignment, the guests labeled by the natural numbers fill up the entire hotel, filling every single positive bed and all the negative beds too!

We end this curious section with yet one more curiosity.

Example 6.15. Let A be a set and let $h : A \to \mathcal{P}(A)$ be any function from A to its powerset. So for each $a \in A$ we have $h(a) \subseteq A$.

Define now another subset of A by $C_h = \{a \in A \mid a \notin h(a)\}$. The curious property of this definition is that, for any element $a \in A$, we must have $C_h \neq h(a)$. This is because the subsets C_h and $h(a)$ disagree over the membership of the element a. Since the subset C_h is not any of the sets $h(a)$, the function h cannot be onto. $\diamond$

You may think this is not much of a curiosity. After all, if $A = \{a, b, c\}$ then $|A| = 3$ and $|\mathcal{P}(A)| = 2^3 = 8$, so the domain is simply too small to support an onto function to the target. So for finite sets, the fact that there is no onto function $A \to \mathcal{P}(A)$ is not a surprise since $|A|$ is always far too small.

But the argument in the example is valid for any domain set, even an infinite set. There is no onto function $\mathbb{N} \to \mathcal{P}(\mathbb{N})$ either. But how can it be true that the set of natural numbers is too small? There is an infinite number of them. There are so many that, in the Hilbert Hotel example, we can fit in all the integers and have plenty of room left over. But there is no way to fill a hotel with beds labeled by $\mathcal{P}(\mathbb{N})$ if you *only* have enough guests to be labeled by $\mathbb{N}$. We are forced to write $|\mathbb{N}| < |\mathcal{P}(\mathbb{N})|$ and to accept the fact that there are different types of infinite sets, and that they are compared by one-to-one and onto functions.

Exercises

1. Define $f : \mathbb{Z} \to \mathcal{P}(\mathbb{Z})$ by $f(n) = \{m \in \mathbb{Z} \mid m \geq n^2\}$.
   Define C_f by $C_f = \{n \in \mathbb{Z} \mid n \notin f(n)\}$ and verify that C_f is not $f(n)$ for any n.
2. Define a function $f : \{a, b, c, d, e, f, g\} \to \mathcal{P}(\{a, b, c, d, e, f, g\})$ so that $C_f = \{a, d, e, g\}$.
3. Here is a "proof" that $\mathcal{P}(\mathbb{N})$ is countably infinite. What's wrong with it?
   We define an one-to-one function $f : \mathcal{P}(\mathbb{N}) \to \mathbb{N}$ as follows. Pick a set in $\mathcal{P}(\mathbb{N})$ at random, A, assign $f(A) = 0$. Continue inductively, at the nth step, choose any set you haven't chosen already, X, and assign $f(X) = n$. There are an infinite number of sets in $\mathcal{P}(\mathbb{N})$, so you don't run out, and the function is one-to-one since you never reuse a number. So $|\mathcal{P}(\mathbb{N})| \leq |\mathbb{N}|$, hence, since we already know $|\mathcal{P}(\mathbb{N})| \geq |\mathbb{N}|$, Cantor is an idiot and $|\mathcal{P}(\mathbb{N})| = |\mathbb{N}|$.

6.6   Cardinality of infinite sets[†]

The moral of the mental experiments of Section 6.5 is that the idea of "number of elements" is too slippery to be useful for infinite sets, it leads to many serious errors. This approach has long been abandoned by mathematicians in favor of the more precise notion of *cardinality* which, while synonymous with the number of elements for finite sets, veers radically away for infinite sets. Infinite cardinals are not *evaluated*, like finite cardinals, where we write $|\{a, b, c\}| = 3$. Mathematicians will not write $|\mathbb{Z}| = \infty$, since ∞ is not a symbol for a cardinal.[1] Instead of being evaluated, infinite cardinals are *compared*, the comparison being done by the existence or non-existence of one-to-one and onto functions.

If there exists a one-to-one function $f : A \to B$, then we write $|A| \leq |B|$, otherwise, if none exists, we say $|B| > |A|$.

Equivalently, if there exists an onto function $f : A \to B$, then we write $|A| \geq |B|$, otherwise, if none exists, we say $|B| < |A|$.

It is true, but not obvious, that the one-to-one version and the onto version are equivalent. If $|A| \leq |B|$ and $|B| \leq |A|$ we write $|A| = |B|$, and say that A and B have equal cardinality. If $|A| = |B|$ then it is true, but again not obvious, that there is a function $f : A \to B$ which is both one-to-one and onto. It is not too hard to prove, most people take only two or three hours to work out all the tricks needed.

---

[1] You may sometimes see the notation $|\mathbb{N}| = \aleph_0$, but we will not use this. Look up "infinite cardinals" if you are curious.

Naively we would expect that all infinite sets have equal cardinality, but this is contradicted by our general cardinality result that $|A| < |\mathcal{P}(A)|$, which is valid even for infinite sets. In fact, there must then be an infinite number of different cardinalities of infinite sets, for instance

$$|\mathbb{N}| < |\mathcal{P}(\mathbb{N})| < |\mathcal{P}(\mathcal{P}(\mathbb{N}))| < |\mathcal{P}(\mathcal{P}(\mathcal{P}(\mathbb{N})))| < \cdots$$

For discrete mathematics, we are most interested in distinguishing the least infinite cardinal from all the rest, and the common terminology of the subject supports this approach. We say that the cardinality of $\mathbb{N}$ is *countably infinite*, and all the other types of infinite cardinals are lumped together under the label *uncountable*. If a set is called *countable*, then its cardinality is either finite or countably infinite. If a set is uncountable, it is often of little interest to distinguish it further.

There are other uncountable sets besides $\mathcal{P}(\mathbb{N})$. From $|A| < |\mathcal{P}(A)|$ we see that the power set of any infinite set is uncountable. It is true, but not obvious, that

$$|\mathcal{P}(\mathbb{N})| = |\mathcal{P}(\mathbb{Z})| = |\mathbb{R}| = |\mathbb{C}|,$$

with the equalities established by producing a one-to-one function of the correct type. It has been unknown for more than a century whether there is any uncountable set with cardinality strictly less than these.

Uncountability is a strange property. By any reasonable measure, "most" infinite sets are uncountable, but most of the sets we actually work with are countable, either finite or countably infinite. So it is natural to presume by familiarity that an unknown set A is countable and write "Let the elements of A be denoted $a_0, a_1, a_2, a_3, \ldots$", however, this may lead to a huge error since, if A is uncountable, the elements of A can have no such naming scheme – $\mathbb{N}$ is merely countable and there wouldn't be enough subscripts!

The set of natural numbers has the smallest infinite cardinal, so $|\mathbb{N}| \le |A|$ for an any infinite set A. For the integers we have $|\mathbb{N}| \le |\mathbb{Z}|$, and both the Hilbert Hotel functions in Section 6.5 show $|\mathbb{Z}| \le |\mathbb{N}|$, so $|\mathbb{N}| = |\mathbb{Z}|$, and the set of integers $\mathbb{Z}$ is also said to be countably infinite.

A surprisingly important example is the set of dots in a square array in the plane, see Fig. 6.4. The set of dots is easily counted (by an onto function from $\mathbb{N}$) via the meandering diagonal path pictured. Since the dots can be labeled with the coordinates (n, m), with $n, m \in \mathbb{N}$, this says that $\mathbb{N} \times \mathbb{N}$ is countable. But perhaps n is just the index of a countable set A and m is the index of a countable set B, so the dot at (n, m) actually refers to the element $(a_n, b_m) \in A \times B$. Now our meandering onto function is establishing that the Cartesian product of any two countable sets is countable. Or perhaps n is the index for a countably infinite collection of countable sets, $A_1, A_2, A_3, \ldots$, and m is the index for counting within each set, so the dot (n, m) refers to the mth element of the nth set, $a_{n,m} \in A_n$. Now our onto function is asserting that a countable union of countable sets is countable.

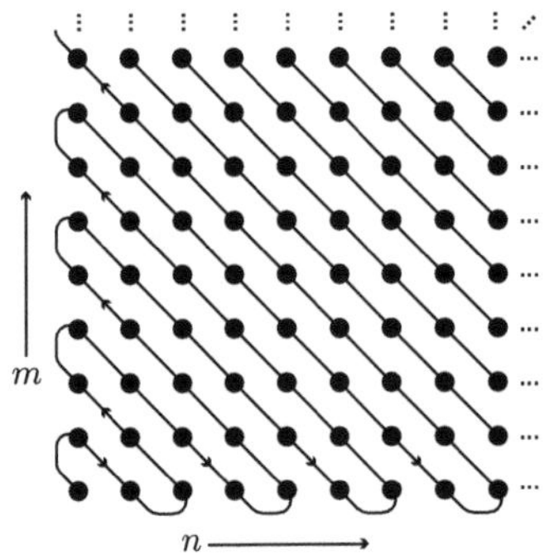

FIGURE 6.4 A countably infinite array of dots.

Let's conclude this section by listing our conclusions about cardinality which relate to infinite sets.

Theorem 6.16 (Cantor's Theorem). *There is no onto function $f : X \to \mathcal{P}(X)$.*

Definition 6.17. We write $|X| \leq |Y|$ if either of the following is true.

- There exists a function $f : X \to Y$ which is onto.
- There exists a function $g : Y \to X$ which is one-to-one.

If $|X| \leq |Y|$ is false, we write $|X| > |Y|$. ♠

Please note that this definition signals a new use of the familiar symbol $\leq$, extending its meaning for ordinary numbers, $1 \leq 17$, or for cardinalities of finite sets $|\mathbb{B}| \leq |\mathbb{D}|$ to the new territory of cardinality of infinite sets. We have seen that it has hidden subtleties and should not be taken for granted.

There is an infinite number of distinct infinite cardinals:

$$|\mathbb{N}| < |\mathcal{P}(\mathbb{N})| < |\mathcal{P}(\mathcal{P}(\mathbb{N}))| < |\mathcal{P}(\mathcal{P}(\mathcal{P}(\mathbb{N})))| < \cdots .$$

All sets A with cardinality $|A| \leq |\mathbb{N}|$ are said to be *countable*. If $|X| > |\mathbb{N}|$, we say X is *uncountable*.

Theorem 6.18 (Useful Countability Results). *The sets $\mathbb{N}$, $\mathbb{Z}$, and $\mathbb{Q}$ are all countably infinite, $|\mathbb{N}| = |\mathbb{Z}| = |\mathbb{Q}|$.*

If $A \subseteq B$ and B is countable, then A is countable.

A subset of a countable set is countable.

If A is countable and $k \in \mathbb{N}$, then $\mathcal{P}_k(A)$ is countable.

A countable union of countable sets is countable. That is, if I is countable, and for each $i \in I$, A_i is countable, then

$$\bigcup_{i \in I} A_i$$

is countable.

If A and B are countable, then $A \times B$ is countable.

Theorem 6.19 (Useful Uncountability Results). *The sets $\mathbb{R}$ and $\mathbb{C}$ are both uncountable.*

If A is an infinite set, then $\mathcal{P}(A)$ is uncountable.

If $A \subseteq B$ and A is uncountable, then B is uncountable.

The set of all infinite sequences of 0's and 1's is uncountable.

Exercises

1. Decide whether $\mathbb{R} \times \mathbb{R}$ is countable or not, and give an explanation. You can use any of the properties we discussed.
2. Decide whether $\mathcal{P}_5(\mathbb{Z})$ is countable or not, and give an explanation. You can use any of the properties we discussed.
3. Consider the set of all real numbers x, with $0 \leq x < 1$), which are expressible as decimals in base 10 such that digits do not decrease as you proceed to the right, like 0.0000011223334444444444444444777777777 .... Decide if this set is countable or not, and give an explanation. You can use any of the properties we discussed.

   Note. The numbers like 0.5 and 0.2224 with invisible zeros at the end are in the set, since $0.5 = 0.49999\ldots$ and $0.2224 = 0.222399999\ldots$.

6.7 Symmetry, reflexivity, transitivity

We started on an examination of set structures, concentrated on relations, and have now spent several sections examining relations of functional type. Recall that in general a relation R is simply a subset of the Cartesian product of two sets. In this section, we will be considering the special case where the two sets are the same.

Let X be a set, a relation on X is a subset $R \subseteq X \times X$. As with ordinary relations, we may use set notation, $(x, x') \in R$, or infix notation $x R x'$ or our diagrammatic notation $x\bullet \longrightarrow \bullet x'$. For these self-relations, the relation diagram may be more compactly rendered. For example the one on the left of Fig. 6.5 contains no more information than the *reduced relation diagram* on the right, in which there is just a single dot for every element of the set X. In the reduced form it is particularly important to not omit the arrowheads. A diagram of this type is also called a *directed graph*. Self-relations in the guise of directed graphs, have been studied extensively, both in their own right, and in the form of flow charts, circuit diagrams, social networks, etc.

In this section we are considering self-relations in the context of three important properties, reflexivity, symmetry, and transitivity.

Definition 6.20. A relation $R \subseteq X \times X$ is said to be *reflexive* if $\forall x \in X; x R x$.

A relation $R \subseteq X \times X$ is said to be *symmetric* if $\forall x, x' \in X; [x R x' \Rightarrow x' R x]$.

A relation $R \subseteq X \times X$ is said to be *transitive* if $\forall x, y, z \in X; [(x R y \wedge y R z) \Rightarrow x R z]$. ♠

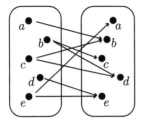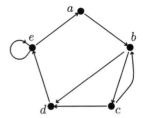

FIGURE 6.5   A relation, and its reduced diagram.

It is important to note the universal quantifications in the definition. Given a relation, that relation *as a whole* is either reflexive, or not, or symmetric, or not, or transitive, or not. One violator is enough to sink the property for the whole relation – for example one element in X with $(x, x) \notin R$ allows you to conclude that R is not reflexive. It is also important to consider these as three separate properties, three qualities that a relation might have.

Reflexivity: This is the easiest one. The definition requires all elements of the set X be related to themselves, or, that each node in the relation diagram has a loop. The relation in Fig. 6.5 is not reflexive since there is no loop at a. The fact that there *is* a loop at d does not change that. Examples of reflexive relations are $\leq$ and $\geq$ for both numbers and cardinality. Also $\subseteq$ on $\mathcal{P}(\mathbb{N})$ and the relation $\Rightarrow$ on Boolean functions defined on the same set are both reflexive.

Symmetry: This property has to be checked for each pair of elements x and x' in X. It requires that either both pairs (x, x') and (x', x) are elements of R, or neither. The relation in Fig. 6.5 is not symmetric because $(a, b) \in R$ but $(b, a) \notin R$. None of the examples of reflexive relations above are symmetric. An example of a symmetric relation on $\{2, 3, 4, \ldots 9\}$ is having no common divisor greater than 1, see Fig. 6.6. For symmetric relations the arrows must occur in matching but reversed pairs, and it makes sense to draw a single straight arrow with two arrowheads as on the left of Fig. 6.6, or, even better, to leave off the arrowheads completely, as on the right.

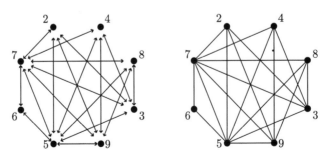

FIGURE 6.6   A symmetric relation and its reduced diagrams.

Symmetry and reflexivity have important anti-versions.

Definition 6.21. A relation $R \subseteq X \times X$ is said to be *anti-reflexive* if $\forall x \in X; \neg x R x$.

A relation $R \subseteq X \times X$ is said to be *anti-symmetric* if

$$\forall x, x' \in X; [(x R x' \wedge x' R x) \Rightarrow x' = x]. \quad \spadesuit$$

Notice the second one, anti-symmetry, in particular. It has a nice familiar form which we will discuss in the next section, but it is not worded as we might expect from the word "anti-symmetric". It does not quite say that it is impossible for both $x R x'$ and $x' R x$ to be true.

Transitivity: This one seems natural in many contexts, but is the most difficult to establish, even with the help of a diagram. It requires, at each $y \in X$, that for every incoming and outgoing pair (x, y), and (y, z) in R there exists a "shortcut" (x, z) in the diagram, connecting x directly to z, see Fig. 6.7, not neglecting situations as on the right where some of the elements coincide.

FIGURE 6.7   Transitivity diagrammatically.

Exercises

1. Consider the relation $\subseteq$ on $\mathcal{P}(\mathbb{N})$.
   Which of the named properties, (reflexive/symmetric/transitive) does it have? Explain.
2. Consider the relation on six elements whose reduced relation diagram is

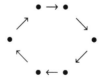

   a) Which of the named properties does it have? Explain.
   b) Suppose that these are just some of the arrows of a transitive relation on six elements. Add all additional arrows to the diagram which are the consequence of transitivity.
3. (The 2 out of 3 problem) Draw reduced relation diagrams for each of the following situations:
   A relation which is reflexive and symmetric, but not transitive.
   A relation which is symmetric and transitive, but not reflexive.
   A relation which is transitive and reflexive, but not symmetric.
   [Hint: diagrams with 3 points work fine.]

6.8 Orderings and equivalence

Many types of relations are defined in terms of reflexivity, symmetry, transitivity, and their anti-versions. It has already been noted that a self-relation is equivalent to a directed graph. If the relation is also symmetric and anti-reflexive, then the relation defines a *simple graph*. These will be considered in more detail later.

A relation which is reflexive, anti-symmetric, and transitive is called a *partial order*. The set on which it is defined is often called a partially ordered set, or a poset for short. Examples of partial order relations are $\leq$, $\geq$, $|$, $\subseteq$, and the implication $\Rightarrow$. At least so far in this text, the least familiar is $|$, or "divides", which defines a partial order the set of positive natural numbers in which we say $n \mid m$ if there exists a $k \in \mathbb{N}$, such that $n \cdot k = m$. So 1 divides all positive naturals, 2 divides all positive even numbers, and 10 divides all positive integers whose decimal representation ends in a 0.

Example 6.22. The relation divides, $|$, is a partial order on the set $X = \{n \in \mathbb{N} \mid n > 0\}$. We just have to check each of the properties. Remember, each property is stated in terms of universal quantification over the ground set, and most involve implications.

Reflexivity: Let $n \in X$ be given. $n = 1 \cdot n$, so $n \mid n$.

Anti-symmetry: Let $n, m \in X$ be given. Suppose $n \mid m$ and $m \mid n$. So $m = kn$ and $n = k'm$. Combining these, we get $n = kk'n$. Since $n > 0$, we may cancel and have $1 = kk'$, so $k = \pm 1$, and since $k \in \mathbb{N}$ we have $n = m$.

Transitivity: Let $x, x', x'' \in X$ be given with $x = kx'$ and $x' = jx''$. So $x'' = j(kx) = (jk)x$, and $x \mid x''$, as required.

It is common for partial order diagrams to omit both the loops and those arrows which are consequences of transitivity, see Fig. 6.8. Perhaps you see why such a relation is called partial order. "Horizontally" the elements are ordered by the arrows, moving from left to right, while "vertically", the elements are not comparable, like all the prime numbers directly related to 1. $\quad \diamond$

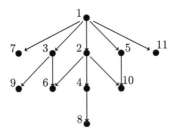

FIGURE 6.8 The partial order divides on $\{1, 2, 3, \ldots, 11\}$.

A completely different type of structure is obtained if we drop "anti-" in anti-symmetry. A relation $E \subseteq X \times X$ is called an *equivalence relation* if it is reflexive, symmetric, and transitive. Equivalence relations occur throughout all of modern mathematics, and working with them is an essential skill in discrete mathematics.

Suppose we have an equivalence relation E on a set X. Let $x \in X$ and set $E_x = \{y \in X \mid yEx\}$ to be the set of all elements related to x. By symmetry, E_x is the same set as the set of all elements to which x is related, $E_x = \{y \in X \mid xEy\}$. By reflexivity, $x \in E_x$, so $E_x \neq \emptyset$. Also, every pair of elements in E_x is related to one another by transitivity: $y, y' \in E_x$ implies yEx ad xEy', so yEy' and by symmetry $y'Ey$ too. So every possible arrow must exist between the elements of E_x. On the other hand, no element of E_x can be related to any element z outside of E_x, since then, by transitivity, x would have to be related to z, violating the definition of E_x. The set E_x is called the *equivalence class* of x. By the above, if x and y are related, then $E_x = E_y$. If x and y are not related, then $E_x \cap E_y = \emptyset$.

Theorem 6.23. *Every equivalence relation on X partitions the elements of X into equivalence classes, $X = \bigcup_{x \in X} E_x$ with $|E_x| \geq 1$ for all $x \in X$, and for each x and $y \in X$, either $E_x = E_y$ or $E_x \cap E_y = \emptyset$.*

So a typical equivalence relation has a diagram like Fig. 6.9, in which there are four equivalence classes. Within each class, all possible relations exist, and between different classes – nothing. Diagrams, if required, for equivalence relations usually omit all the arrows and instead just focus on the separation into equivalence classes.

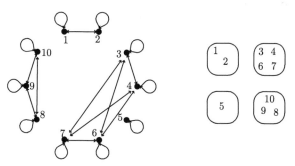

FIGURE 6.9   An equivalence relation on $\{1, 2, 3, \ldots, 10\}$ and the corresponding equivalence classes.

Equivalence relations are very common and are often denoted by symbols such as $\equiv, \sim, \simeq, \asymp, \approx, \cong, \doteq, \Leftrightarrow, \leftrightarrow, \leftrightarrows$, etc. The same pair of objects may be equivalent in one sense and non-equivalent in another, or equivalent in several different senses. So, for example, two bishops b and b' in chess might be equivalent in the sense they are both black, $b \cong b'$, or that they both move on black squares, $b \doteq b'$, or that they are among the bishops, $b \asymp b'$. Analyzing or programming a chess game probably involves several equivalence relations among the pieces.

Exercises

1. Draw the reduced relation diagram of an equivalence relation with equivalence classes $\{a, b, c\}$, $\{e\}$, and $\{d, f\}$.
   For any equivalence relation on $\{a, b, c, d, e, f\}$, how many arrows can it have? List all possibilities.
2. Define a relation $\propto$ on $\mathbb{Z}$ by setting $n \propto m$ if there is a $k \in \mathbb{Z}$ with $n - m = 10k$.
   Show that $\propto$ is an equivalence relation and describe the equivalence classes.
3. Define a relation $\succ$ on $\mathbb{Z}$ by setting $n \succ m$ if there is a $k \in \mathbb{Z}$ with $n + m = 10k$.
   Decide whether or not $\succ$ is an equivalence relation. If it is, give the equivalence classes. If not, show why not.

6.9   Case study: The developer's problem

A developer has a plot of land that he wants to split up for houses. It is a 6×6 grid which he wants to divide into 1×2 blocks, around which houses are to be built. Between every pair of blocks, there will be a road. The current design, Fig. 6.10a has a long straight east/west road cutting through the development, which the developer does not like. Through-roads become busy, and people pay less for houses on busy roads. He has a design for an 8×8 development with no through-roads, and would like a design for a 6×6 grid with the same property.

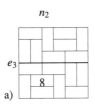
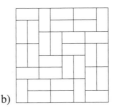

FIGURE 6.10   Can we improve the design?

This is a geometric puzzle, and it does not hurt to take out a set of dominoes and try to arrange them in a 6×6 grid with no through-roads. It might take you just a few minutes to succeed. But what if you don't. Maybe the solution is very tricky, or maybe it cannot be done. Either way, it is time to analyze the problem mathematically.

We have plenty of sets which look promising. There is the set of points in the plane, $\mathbb{R}^2$, the set 10 of possible through-roads, call them $R = \{e_1, e_2, e_3, e_4, e_5, n_1, n_2, n_3, n_4, n_5\}$, the set of $6^2 = 36$ squares in the grid which are naturally arranged as a Cartesian product, $\{1, 2, 3, 4, 5, 6\}^2$, the set B of 18 blocks with which the grid is to be tiled. We have many relationships we can model.

For instance, the 36 squares are not occurring in isolation. Some pairs share a side and so could be made into a block. Other pairs do not and can not. This relation of *adjacency* is obviously important to the problem.

However, let's focus on one particular relation, that of *obstruction*. Once you have designed an arrangement of blocks, some roads become obstructed by some blocks, since the road can only go around, and not pass though the block. In the diagram above, the blocks have been placed so that block 8 obstructs road n_2. Each block arrangement comes with an obstruction relation

$$O = \{(b, r) \in B \times R \mid \text{the line of the road } r \text{ cuts through block } b\}.$$

Is the relation O functional? And, if it is, we certainly want to know if it is one-to-one, or if it is onto.

Functionality would mean that the relation has the property that each block obstructs exactly one road. Since the blocks are 1×2, they cannot obstruct those roads running parallel to their alignment, and each can and must obstruct exactly one road running perpendicular to its alignment. But, you may say, the block may be somewhere in the middle of the design, so it doesn't obstruct anything because the road is obstructed before it ever gets close to the block. To resolve this question we have to use the actual mathematical definition of the relation O, not our general feeling for the word obstruct. In the definition, the road r is related to all blocks its line cuts through, not just the ones it encounters first. So the relation O is functional. Notice that if we had flipped the Cartesian product to $R \times B$, that obstruction relation would not be functional.

One to one: This would mean that each road can be obstructed by at most one block, but we have already seen in the example design that road n_2 is obstructed by 2 blocks. In fact, there are 10 roads and 18 blocks, so it is impossible for any obstruction function to be one-to-one.

Onto: Onto would require that each road is obstructed by at least one block. That is not always true either, since road e_3 in the sample development is not obstructed. So the design need not be onto. But wait – before we move on, we should pause to notice that this is related to the problem which we want to solve. We don't want to show that every obstruction function is onto. We want to find one that is onto, a development that obstructs every road. So we haven't solved the problem, but we have put it on a mathematical foundation:

↬ Is there a development so that the *obstruction function* is *onto*? ↫

The domain of O has 18 elements, and the target has 10, so from that point of view an onto obstruction function looks plausible.

6.10 Case study: Wolf-Goat-Cabbage II

Let's consider the Wolf-Goat-Cabbage problem of Sections 1.1 and 3.8 again from the point of view of relations.

Our first analysis of the problem mathematically was to encode the 2^4 *states* of the problem as one of the elements of $\mathcal{P}(\{M, W, G, C\})$, the elements of the subset recording which objects are on the initial side of the fjord. So $\{M, W, G, C\}$ is the initial state, and the solution state is $\emptyset$.

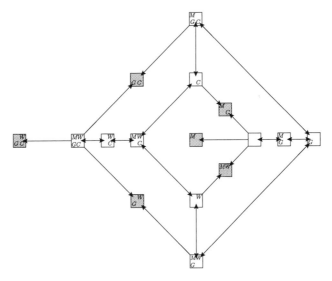

FIGURE 6.11   The relation diagram of Wolf-Goat-Cabbage.

Now we want to define a relation which encodes when two states are connected by the man starting in a stable state and traveling across the fjord exactly once. If he traverses the fjord alone, then the new state is obtained from the old state by either adding or removing M from the state. If he takes object X, then that object also is added or removed from the state set. We call this the *traversing relation* and use the symbol $\rightsquigarrow$. If (A, B) is in the traversing relation set, we write $A \rightsquigarrow B$.

So starting from the state $\{M, W, G, C\}$ we have the sequence of related pairs

$$\{M, W, G, C\} \rightsquigarrow \{W, C\} \rightsquigarrow \{M, W, C\} \rightsquigarrow \{C\} \rightsquigarrow \{M, C\}$$

taking us to state $\{M, C\}$ where we must stop, because $\{M, C\}$ is not a stable state, since the Wolf and the Goat are alone on the opposite shore, and no moves are defined for unstable states. The complete relation diagram for the traversing relation is shown in Fig. 6.11 where the reader should see the huge advantage of relation diagrams. So much of the hidden structure of this problem is revealed first by encoding the problem as a relation, and second by viewing the relation diagrammatically. It is now not only easy to solve the problem, but also possible to study and compare different solutions.

Here is a question you can now answer. Suppose someone advises the Viking to adopt the following strategy. "Only move to a stable state, and, if there is a choice of stable states, just choose the one you prefer". Will this result in the

Viking always finding a solution and crossing the river?

✠

Unfortunately, a viking who really loved his goat and following this strategy could be caught up in an infinite loop like

$$\{M, W, G, C\} \rightsquigarrow \{W, C\} \rightsquigarrow \{M, W, G, C\} \rightsquigarrow \{W, C\}$$
$$\rightsquigarrow \{M, W, G, C\} \cdots$$

in which he rows the goat back and forth and never does anything else. To avoid this, the strategist adds the rule "Never perform the same action twice in a row". Now must the Viking win?

6.11   Case study: The non-transitive dice

A important relationship is "being better than". You have certainly seen many contests and playoffs which purport to decide which player or team is the best. If we have a set of teams, the relationship we want is $x \succ y$, that team x beats team y in the playoffs. We should expect the relation $\succ$ to be a strict partial order, that is, to be anti-reflexive, anti-symmetric, and transitive. Anti-reflexive since no team beats itself, anti-symmetric since the game is played so there is a clear winner and no tie, and transitive since the skill level should determine the outcome.

So lets say we design a playoff among just four teams, Westfield, Schenectady, Rheinbeck, and Lake Placid, i.e., W, S, R, and P. Suppose on the first day of the playoff that two games are played, with results $P \succ R$ and $W \succ S$. To find the winner, we need a playoff of P versus W, and to identify the worst team, we can also play R versus S at a different venue. That determines the overall winner, but will leave open the question of how to rank the middle two teams. If that is important, perhaps because it will affect next year's recruiting, you can schedule the last pair for a game on day three. But, with the championship already decided, that will not be a popular game. It will be poorly attended and there may even be a conflict as to whether the contestants want to win or lose such a game.

So the organizers decide instead, for the second day, to have a match pitting the two winners of the first day against the two losers they haven't played yet; P vs. S and W vs. R. If both winners win again, then an exciting third day-three match-up is required between the two clear winners, and the two clear losers.

On the other hand, if just one of the previous winners wins, say $P \succ S$, but $R \succ W$, then we know that, since $P \succ R \succ W$, we must have $P \succ W$ by transitivity, so P wins the playoff. In this case we have also ranked the teams since $P \succ R \succ W \succ S$. In this case the playoff is all done in two days. The only odd thing for those who follow sports is that P has emerged victorious on day two by winning against the worst team in the league, not against the second ranked team.

The only case left is if both winners actually lose on day two. Again, those who watch sports would not be overly surprised, since there is always some element of chance. Nevertheless it would be regarded as a fluke, an unexpected result, something outside the game, if the results so far looked like

$$
\begin{array}{ccc}
P & \prec & S \\
\curlyvee & & \curlywedge \\
R & \succ & W
\end{array}
$$

Such a configuration of results violates our presumption about how a game of skill should behave, since it violates transitivity and anti-symmetry. Given the game results $P \succ R \succ W$, transitivity requires $P \succ W$, and transitivity with $W \succ S \succ P$ requires $W \succ P$. But if $P \succ W$ and $W \succ P$ then anti-symmetry requires $P = W$, which is false.

If this should happen, the commentators will be happy to explain how the complexities of the game, the rules, the penalties and the configurations of the teams involved, all contribute to make this possible. But should we believe them?

Here is a very simple game that does not take any more than the multiplicative principle to analyze. Instead of teams there are four dice constructed as in Fig. 6.12. The dice are rolled and the higher score wins. The dice are a little unusual, but they look fair in that each die has 24 pips. The 24 pips are distributed so that no two dice have the same numbers, hence no ties are possible. There are 6^2 possible outcomes for rolling two dice, and if for more than half that number, die a beats die b, we write $a \succ b$.

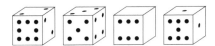

FIGURE 6.12   Magic Dice: $\{2, 2, 2, 2, 8, 8\}$, $\{3, 3, 4, 4, 5, 5\}$, $\{0, 0, 6, 6, 6, 6\}$, $\{1, 1, 1, 7, 7, 7\}$.

We might expect that such a simple match-up must have results in a strict partial order, but it is easy to check that

$$
\begin{array}{ccc}
\{2, 2, 2, 2, 8, 8\} & \succ & \{1, 1, 1, 7, 7, 7\} \\
\curlywedge & & \curlyvee \\
\{3, 3, 4, 4, 5, 5\} & \prec & \{0, 0, 6, 6, 6, 6\}
\end{array}
$$

with 24 wins to 12 losses in each victory. So the relationship cannot be a strict partial order, even in this simple contest.

And we are not talking about chance occurrences either, since the ordering with its non-transitivity is defined by the most likely outcome in each case. Even with dice whose numbers seem to have only two aspects; the extreme values on the die versus how often those extreme values are represented; we have non-transitivity. Just those two aspects are enough to explode the notion of "skill level" which makes the strict partial order seem to be the obvious relation type to expect for sport tournaments. 'Skill level' is one of those ideas which leads us to expect a one-dimensional result, but even for our simple dice we find a multi-dimensional phenomenon.

So even in something so direct and pure as four sailors on a submarine contesting by arm wrestling, there are enough aspects, bicep strength versus arm length, to make the contest results interesting, complex, and perhaps non-transitive.

6.12 Case study: The developer's problem II

Let us return to the developer's problem of Section 6.9. We want to find an arrangement of blocks on a 6×6 grid so that the obstruction function O is onto. There are plenty of blocks to obstruct the roads, but the issue is that, in most of the easy examples to construct, many blocks seem to be obstructing the same road. So let's define another relation to cover the situation of a block obstructing a road already obstructed by another block.

$$E = \{(b, b') \in B \times B \mid [\exists r \in R; ((b, r) \in O) \wedge ((b', r) \in O)]\}.$$

In other words, some road obstructed by b is also obstructed by b'.

This can be a very tricky relation for blocks of arbitrary shape, but because they are all 1×2, the obstruction relation is actually an equivalence relation.

Reflexivity: Each block obstructs a road, so each block is related to itself.

Symmetry: This is obvious from the commutativity of $\wedge$ in the definition.

Transitivity: If blocks i and j obstruct road r, and blocks j and k obstruct road r', then block j only obstructs one road, so $r = r'$ and so $(i, k) \in E$, as required.

Here are the six equivalence classes for another development design.

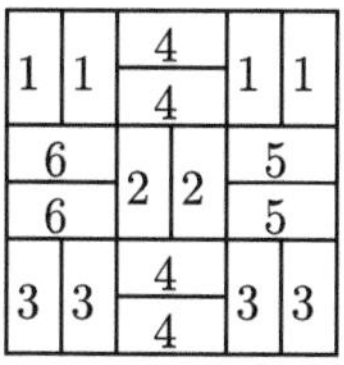

1	1	4	1	1
		4		
6		2 2	5	
6			5	
3	3	4	3	3
		4		

FIGURE 6.13 Obstruction classes.

Notice in Fig. 6.13 that each equivalence class has even cardinality. Let prove that true in general by induction. You might ask, what is the "for all n" statement that we want to prove? The induction will be very short. We will show that each of the 5 north/south roads is obstructed by an even number of blocks.

Base case: The leftmost road. The first column of squares is filled partially by some north/south aligned blocks, which uses up an even number of squares, and partially by some east/west aligned blocked, which stick out and obstruct road one. Since the north/south aligned used up an even number of the six squares in column one, road one is obstructed by an even number of blocks.

Inductive step. Suppose the nth road is obstructed by an even number of blocks. Consider the column of squares on its right edge. An even number of squares are used up by the north/south aligned blocks, and an even number of stick out to the west obstructing road n by the inductive hypothesis, so there are an even number left to stick out to the right and obstruct road $n + 1$.

So all 5 north/south roads are obstructed by an even number of roads by induction. The same finite inductive argument works for the five east/west roads. This completes the proof.

Now let's define a new obstruction function O' whose domain is the set of equivalence classes, to the set of roads, assigning each class to the road which each element in the class obstructs. Again, we want to know if it is possible for O' to be onto, and obstruct all the roads.

How many equivalence classes can there be? Equivalence classes are never empty, and these are of even cardinality, so the 18 blocks can be in at most 9 equivalence classes. But there are 10 roads. So the new obstruction function O' can never be onto for any design, and so it is impossible to obstruct all 10 roads.

Observe that in Fig. 6.13 the blocks are partitioned into 6 equivalence classes, so four roads are unblocked. Can you find them? In a sense, the design in Fig. 6.10 is optimal since there is only one unblocked road.

Note that for the 8×8 grid we have, by the same arguments, 16 possible equivalence classes and only 14 roads to block. Identify the 14 equivalence classes for the 8×8 design in Fig. 6.10.

6.13 Case study: The missing region problem II

In Section 2.8 you read about a circle C in the plane bounding a disc D, a finite subset $V \subseteq C$, all the $\binom{|V|}{2}$ chords joining the elements of V, and the set of regions R into which they cut the circle. The problem was to find $|R|$.

In this section we want to define a relation $\Psi \subseteq (\mathcal{P}_0(V) \cup \mathcal{P}_2(V) \cup \mathcal{P}_4(V)) \times R$ between certain subsets of V and the set of regions R.

Definition 6.24. A set $X \in \mathcal{P}_0(V) \cup \mathcal{P}_2(V) \cup \mathcal{P}_4(V)$ is related to region $r \in R$ by Ψ, or $X \Psi r$, if one of the following holds:

- $X \in \mathcal{P}_4(V)$ and the lowest point of r is both an interior point of the circle, and the intersection of two chords whose endpoints are in X.
- $X \in \mathcal{P}_2(V)$, the lowest point of r is an element of X, the chord c joining the two elements of X contains more than one point of the boundary of r, and the chord c separates r from the lowest point of the circle.
- X is empty, $X \in \mathcal{P}_0(V)$, and r contains the lowest point the circle. ♠

The relation Ψ can be a bit quirky in some special situations. So, if necessary, turn the figure slightly so that the lowest point of the circle is not an element of V, and so that no chord is either horizontal or vertical. Also, we know that we don't get the maximum number of regions in situations like the regular hexagon, where all the chords meet in the middle, so let's only consider the situation where at most two chords meet at any interior point. Call this a *general* collection of chords, see Fig. 6.14.

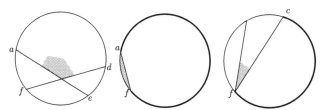

FIGURE 6.14   Illustrating the relation Ψ for non-empty sets of a general collection. The region r is shaded.

Theorem 6.25. *For a general figure, the relation Ψ is functional.*

Proof. If $X \in \mathcal{P}_4(V)$, and the elements of X are labeled a, b, c, and d as you go clockwise around the circle, then only chords ac and bd intersect in the interior of the circle, and since the figure is general, exactly four regions meet there, and for exactly one is the intersection point the lowest point of the region.

If $X \in \mathcal{P}_2(V)$, then its elements a and b, with b lower than a, define a chord. The chord is divided into segments by the other chords, and the segment at b separates exactly two regions, one on the side of the lowest point of the circle, and one on the other side, so X is related to exactly one region of r.

Since the figure is general, the lowest point on the circle is not in V, and $\emptyset$ is related to exactly one region.   □

So, for general figures, the relation Ψ is functional, $\Psi : \mathcal{P}_0(V) \cup \mathcal{P}_2(V) \cup \mathcal{P}_4(V) \to R$ and we can write $\Psi(X)$. The function Ψ is evaluated in Fig. 6.15 for an example.

Of course, you must suspect what comes next:

Theorem 6.26. *For general figures, the function Ψ is one-to-one and onto.*

Proof. Ψ is onto: Let r be a region. It has a lowest point. If that lowest point is an interior point of the circle, then it is the intersection of two chords, which have four endpoints $\{a, b, c, d\}$, and $\Psi(\{a, b, c, d\}) = r$. On the other hand, if the lowest point is on the boundary, that lowest point is either an element of V, or it is the very lowest point of the circle. If it is an element $b \in V$, then either one or two chords at b are boundaries of r, and one of them, ab, separates r from the lowest point, so $\Psi(\{a, b\}) = r$. Lastly, if the lowest point of r is the lowest point of the circle, then $\Psi(\emptyset) = r$.

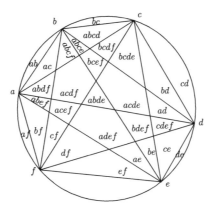

FIGURE 6.15 The relation Ψ for 6 boundary points.

Ψ is one-to-one: Suppose $\Psi(X) = \Psi(Y) = r$. Since for a general figure each region can have only one lowest point, and it is either on the boundary or in the interior, we must have $|X| = |Y|$. If $|X| = |Y| = 4$, then the two chords meeting at the lowest point of r have only 4 endpoints, so $X = Y$. If $|X| = |Y| = 2$, and the lowest point b of r is an element of both sets, and since there is just one chord at b separating r from the lowest point of the circle, the oppositive point of the chord must be in both sets as well, so $X = Y$. Lastly, if $|X| = |Y| = 0$, then $X = Y = \emptyset$. $\qquad\square$

Now that we have a one-to-one and onto function we know that

$$|R| = |\mathcal{P}_0(V) \cup \mathcal{P}_2(V) \cup \mathcal{P}_4(V)| = \binom{|V|}{0} + \binom{|V|}{2} + \binom{|V|}{4} \qquad (6.5)$$

and this explains exactly the missing region problem. For $|V| \le 5$, the regions of a general figure correspond exactly to the even subsets of the set V, which are half of all the subsets and given by $2^{|V|-1}$. So the number of regions advances geometrically for the first five terms. After that, for $|V| \ge 6$, the regions correspond only to the small even subsets of $|V|$, those of cardinality at most 4, and the geometric progression is broken.

We also see from Eq. (6.5) that the number of regions of a general figure does not grow exponentially with the cardinality of V, but only as a 4th degree polynomial. As a side benefit, we have an efficient method for labeling the regions inside a general figure.

6.14 Summary exercises

You should have learned about:

- The definition of a binary relation.
- How to create and interpret a relation diagram.

- The abstract definition of a function.
- The precise formulations of one-to-one and onto.
- The connection between one-to-one, onto, and cardinality for finite sets.
- How to count the number of functions of various types between finite sets.
- How to show a function is one-to-one or onto.
- How infinite cardinals are compared by one-to-one or onto functions[†].
- The distinction between Countable and Uncountable sets[†].
- Basic properties of Countable sets[†].
- Basic properties of Uncountable sets[†].
- The reduced diagrams for self-relations.
- The meaning of reflexivity, anti-reflexivity, symmetry, anti-symmetry, and transitivity.
- The definition of a partially ordered set.
- The Equivalence Relation and its structure.
- How to determine equivalence classes.

1. a) Define a function $f : \mathcal{P}_2(\mathbb{N}) \to \mathcal{P}_f(\mathbb{N})$ which is one-to-one.
   b) Define a function $g : \mathcal{P}_f(\mathbb{N}) \to \mathcal{P}_2(\mathbb{N})$ which is one-to-one. In each case, draw the correct conclusion about cardinalities.

2. How many relations are there from set $\{a, b, c\}$ to set $\{1, 2, 3\}$? How many are of functional type? How many are one-to-one functions? How many are one-to-one and onto functions?

3. Let $X = \{a, b, c\}$ and consider the relation $\subset$ on $\mathcal{P}(X) \times \mathcal{P}(X)$ defined by $A \subset B$ if $A \subseteq B$ and $A \neq B$. Draw the relation diagram.
   Is $\subset$ of functional type? Why or why not?

4. How many relations on the finite set X have at least one element related to itself?
   [Hint: Use inclusion/exclusion.]

5. Define a function $f : \mathbb{N} \to \mathcal{P}(\mathbb{N})$ By setting $f(n)$ to be the set of all numbers which occur as consecutive strings in the decimal representation of n.
   Example: $f(5280) = \{0, 2, 5, 8, 52, 28, 80, 528, 280, 5280\}$.
   Is f one-to-one or not? If so, prove it. If not, provide a counterexample.
   Is f onto or not? If so, prove it. If not, provide a counterexample.
   What do these functions imply about the cardinality of $\mathcal{P}(\mathbb{N})$

6. Define a one to one function with domain $\{a, b, c, d, e, f\}$ and target $\mathcal{P}(\{a, b, c, d, e, f\})$.

7. Find a one-to-one and onto function from $\mathcal{P}_2(\{1, 2, 3, 4, 5\})$ to $\mathcal{P}_3(\{1, 2, 3, 4, 5, 6\})$, or show that one does not exist.

8. Find an onto function from $\mathcal{P}_2(\{1, 2, 3, 4, 5, 6\})$ to $\mathcal{P}(\{2, 4, 6\})$, or show that one does not exist.

9. Let X be a finite set and define $f : X \to \mathcal{P}(\mathcal{P}(X))$ by setting

$$f(x) = \{A \subseteq X \mid x \in A\}$$

Show whether or not f is one-to-one.

Show whether or not f is onto.

10. Show that for any 16 subsets of 0,1,2,3,4,5,6 with at most 3 elements, there are at least two of these subsets with equal sums of elements, like $\{1, 3, 5\}$ and $\{2, 3, 4\}$.

11. Let $A = \{1, 2, 3, 4, 5\}$ and $B = \{1, 2, 3, 4, 5, 6, 7, 8\}$. How many onto functions are there from A to B?

    How many onto functions are there from B to A?

12. Our multiple Cartesian products always result in tuples with a finite number of coordinates. What if we wanted an infinite number, like this:

$$M = \bigtimes_{n=0}^{\infty} A_n$$

What if the A_n's were all countable. Must M be countable?

Chapter 7

Elementary number theory

7.1 Primality, the Sieve of Eratosthenes

If you've ever read an account of the origin of numbers, it probably went something like this: Prehistoric people, "cavemen", invented the natural numbers ($\mathbb{N}$) for counting their goats. As time passed, their primitive idea of number was expanded, adding negatives ($\mathbb{Z}$) for accounting, fractions ($\mathbb{Q}$) for proportions, then the rest of the real numbers ($\mathbb{R}$) for geometry, and later complex ($\mathbb{C}$) and even more exotic numbers used in modern engineering and physics. Accepting this story of unimpeded mathematical progress, we should no more want to go back and restrict ourselves to those primitive numbers than we would want to walk around wearing bear skins.

But our study of discrete mathematics tells us that we should do just that, since many important discrete problems do involve just counting and integers. Early in the text we considered $\mathbb{Z}$ merely as a set of interesting objects. In number theory, the arithmetic of the integers plays the major role, specifically addition and multiplication.

You should know from algebra that addition and multiplication are both commutative; $i + j = j + i$, $ij = ji$, associative $i + (j + k) = (i + j) + k$, $(ij)k = i(jk)$; that multiplication distributes over addition, $i(j+k) = ij + ik$; 0 is the additive unit $i + 0 = i$ and 1 is the multiplicative unit, $1j = j$; and finally that every integer i has an additive inverse $-i$, with $i + (-i) = 0$.

The presence of the additive inverse makes subtraction possible.

Multiplicative inverses in $\mathbb{Z}$, however, are very rare. In $\mathbb{Z}$, only 1 and -1 have multiplicative inverses: $1 \cdot 1 = (-1)(-1) = 1$. So division in $\mathbb{Z}$ requires special attention.

Definition 7.1. *Division with remainder* is an integer equation $n = mq + r$ with $0 \leq r < m$. The number q is the *quotient* of n with respect to m, and r is called the *remainder*. If $r = 0$, then we say m *divides* n, and write $m \mid n$. ♠

If $p \geq 2$, and its only divisors are 1 and p, then p is a *prime*. The set of all primes is denoted by $\mathbb{P}$, as mentioned previously.

Given a number n, the most obvious way to determine whether it is prime, $n \in \mathbb{P}$, is to check whether $m \mid n$ for all values m in the range $2 \leq m < p$. There are two ways to improve this naive approach. The first is to note that you only have to check n for prime divisors. The second is to note that if $n = ab$, then a and b cannot both be greater than $\sqrt{n}$. This gives us

Theorem 7.2. *If $n > 1$ is not prime, then n has a prime divisor no larger than the square root of n.*

Theorem 7.2 is the basis for a method of determining primes known as the *Sieve of Eratosthenes*. You start with a list of all primes up to N, $P_1 = \{2, 3, 5, \dots\}$ and then form a second list L of all values in $\mathbb{N}$ up to N^2. Then cross off from L the values 0, 1 and all multiples of elements in P_1. When that is done, all remaining elements of L are added to P_1 to form P_2, which by Theorem 7.2 contains all the primes up to N^2.

Example 7.3. If $P_1 = \{2, 3, 5, 7\}$, which are all the primes up to $N = 10$, then, indicating the crossed out terms with subscripts, the sieved list L

0_x	1_x	2_2	3_3	4_2	5_5	$6_{2,3}$	7_7	8_2	9_3
$10_{2,5}$	**11**	$12_{2,3}$	**13**	$14_{2,7}$	$15_{3,5}$	16_2	**17**	$18_{2,3}$	**19**
$20_{2,5}$	$21_{3,7}$	22_2	**23**	$24_{2,3}$	25_5	26_2	27_3	$28_{2,3,7}$	**29**
$30_{2,3,5}$	**31**	32_2	33_3	34_2	$35_{5,7}$	$36_{2,3}$	**37**	38_2	39_3
$40_{2,5}$	**41**	$42_{2,3,7}$	**43**	44_2	$45_{3,5}$	46_2	**47**	$48_{2,3}$	49_7
$50_{2,5}$	51_3	52_2	**53**	$54_{2,3}$	55_5	$56_{2,7}$	57_3	$58_{2,3}$	**59**
$60_{2,3,5}$	**61**	62_2	$63_{3,7}$	64_2	65_5	$66_{2,3}$	**67**	68_2	69_3
$70_{2,5,7}$	**71**	$72_{2,3}$	**73**	74_2	$75_{3,5}$	76_2	77_7	$78_{2,3}$	**79**
$80_{2,5}$	81_3	82_2	**83**	$84_{2,3,7}$	85_5	86_2	87_3	$88_{2,3}$	**89**
$90_{2,3,5}$	91_7	92_2	93_3	94_2	95_5	$96_{2,3}$	**97**	$98_{2,7}$	99_3

records all the primes $P_2 = \{2, 3, 5, 7, 11, 13, 17, 19, 23, 29, 31, 37, 41, 43, 47, 53, 59, 61, 67, 71, 73, 79, 83, 89, 97\}$ up to $10^2 = 100$. $\diamond$

We could now use the list in the example to sieve out a list of the primes up to $100^2 = 10,000$. In doing so, we certainly would use a computer, but historically the automation has been done also with various mechanical methods, such as strips of paper with appropriately punched holes – 'the sieves'.

Example 7.4. The 13^2 problem. Show that $13^2 \neq 11 \cdot 19$ using only the primality of 11, 13, and 19.

Of course, you can multiply, $13^2 = 169$ and $11 \cdot 19 = 209$, but that is "cheating". The solution should look more like what you naturally will do to show that $5^2 \neq 2 \cdot 13$, where you don't multiply at all and just note that the number on the left is odd, and the number on the right is even. So we are looking for an argument that only uses the fact that 11, 13, and 19 are primes. $\diamond$

Primality arguments involve the unique factorization of an integer into primes, and the 13^2 problem highlights that this fact is not obvious.

1. Use the Sieve of Eratosthenes to find all the primes up to 166.
2. Suppose you wanted to decide if 507 was prime. What is the largest prime you need to check?

   Is 507 prime? If not, give a prime factorization.
3. Give a prime factorization of 1776. Begin as efficiently as possible.

7.2 Common divisors, the Euclidean Algorithm

It might seem that Theorem 7.2 made the determination of primality very easy. After all, to determine if a number on the order of 10,000 is prime, we only have to check at most $\sqrt{10,000} = 100$ factors – and half of those are even numbers, for all of which only the factor 2 needs to be checked.

But we should not be too happy about the fact that 100 is so much less than 10,000. Our benchmark for dealing with numbers must take into account that we have a very efficient way of encoding numbers using, say, the binary or decimal number system. When we do arithmetic, adding and multiplying, we do that by working with the digits or the bits alone, not the number itself. To add two N-bit numbers persuades us to do N bit-additions, and then handle at most N carries. Ignoring details of implementation, we expect to do about $2N$ little tasks. For multiplication there is more work to do. We gather the results of multiplying each bit of one factor by all the bits of the other factor, and adding the shifted results. So we expect, say, $2N^2$ little tasks. That sounds like an awful lot until you realize that the numbers we are talking about, N-bit numbers, are of size roughly 2^N.

Now consider, for an N-bit number, if we want to use Theorem 7.2 and just check primality by looking for small factors, we are talking about $\sqrt{2^N} = (\sqrt{2})^N$ divisions, an amount of work which is exponential in the number of bits, and that is very bad for scaling. What we would want is a method in which the work required is proportional to, or even a polynomial in, the number of bits. No test for divisors comes close to that. Not so for *common divisors*.

Definition 7.5. Given integers $d, n, m > 0$, the integer d is said to be a *common divisor* of n and m if $d \mid n$ and $d \mid m$.

The *greatest common divisor* of n and m, written $\gcd(n, m)$, is a common divisor with the property that $d \leq \gcd(n, m)$ for all common divisors d of n and m. ♠

If $\gcd(n, m) = 1$, then n and m are said to be *relatively prime*, or *coprime*.

Example 7.6. Let's compute $\gcd(1776, 5280)$.

✠

Factoring, $1776 = 2 \cdot 888 = 2^4 \cdot 111 = 2^4 \cdot 3 \cdot 37$, and $5280 = 8 \cdot 660 = 8 \cdot 6 \cdot 11 \cdot 10 = 2^5 \cdot 3 \cdot 5 \cdot 11$. So $\gcd(1776, 5280) = 2^4 \cdot 3 = 48$. ◇

The solution of the previous example used *unique prime factorization*, which we have only hinted at so far, but even if that is correct, it is not a good general method. We just saw that, if you want to factor a number into its prime factors, just expect to do a lot of work.

Theorem 7.7. Let $n = m \cdot q + r$ be division with remainder of n by m with quotient q and remainder $r > 0$. Then $\gcd(n, m) = \gcd(m, r)$.

Proof. We will show that, in fact, the pair $\{n, m\}$ has the same set of common divisors as the pair $\{m, r\}$, and we'll use the double inclusion method.

Let d be a common divisor of m and r, so $m = id$ and $r = jd$. Then $n = mq + r = idq + jd = (iq + j)d$, and $d \mid n$, and so d is a common divisor of n and m.

On the other hand, let d' be a common divisor of n and m, so $n = ud'$ and $m = vd'$. Then $ud' = vd'q + r$, so $r = ud' - vd'q = (u - vq)d'$, and so d' is a common divisor of m and r. □

Example 7.8. To compute $\gcd(1776, 5280)$ using Theorem 7.7, divide $5280 = 1776 \cdot 2 + 1728$ giving $\gcd(5280, 1776) = \gcd(1776, 1728)$. But don't stop there. Divide again $1776 = 1728 \cdot 1 + 48$ giving $\gcd(1776, 1728) = \gcd(1728, 48)$. Then $1728 = 48 \cdot 36 + 0$, so $48 \mid 1728$, and of course $\gcd(1728, 48) = 48$. And that means $\gcd(1776, 5280) = 48$. ◇

This repeated use of Theorem 7.7, as in Example 7.8, is called the Euclidean Algorithm.

Algorithm 7.9 (Euclidean Algorithm). To compute $\gcd(n_0, n_1)$ with $n_0 > n_1 > 0$; in the kth step, $k \geq 0$, perform division with remainder $n_k = n_{k+1} \cdot q_{k+1} + n_{k+2}$ to define the new value n_{k+2}. The procedure halts at the Kth step when $n_{K+2} = 0$. Then $\gcd(n_0, n_1) = n_{K+1}$. ♡

So which do we use? The direct method of Example 7.6 or the Euclidean Algorithm as in Example 7.8? Nobody likes to divide, so what is the worst case for the Euclidean Algorithm? In division with remainder $n = m \cdot q + r$ there are two possibilities, either m is small, $m \leq n/2$, in which case the remainder which is yet smaller than m, satisfies $r < n/2$; or m is large, $n/2 < m < n$, in which case division with remainder removes just one m from n and $r = n - m < n/2$. In either case, $r < n/2$, or in terms of the binary numbers, one bit fewer.

For the Euclidean Algorithm, this means that n_{k+2} has at least one bit fewer than n_k, and an N-bit number has at most $2N$ divisions required. In the example, 5280 is a 13-bit number, we would have felt unlucky indeed had it taken 26 divisions, but even then, the $2N$ is proportional to the number of bits, and is the clear winner over $(\sqrt{2})^N$ for large examples, and for most small examples, too.

Exercises

1. Use the Euclidean Algorithm to find the greatest common divisor of 123 and 321.
2. Use the Euclidean Algorithm to find $\gcd(988887, 888887)$. Take any shortcuts you like. (But only hand calculations.)
3. Show using the Euclidean algorithm that 100 and 169 are coprime.

7.3 Extended Euclidean Algorithm

Example 7.10 (Euclid's Coin Problem). Suppose Alexander the Great orders his treasury to make only two types of coin, one with his own picture, the A coin, and a less valuable one with the picture of the only other figure of importance in his empire, his horse Bucephalus, the B coin. They are to be the only coins in the land, and they must be used to make change for any amount of Talents. What whole number values of Talents can A and B have?   ◇

If $A = 16$ and $B = 1$ then the system clearly works, and any system with $B = 1$ works since you really don't need the A coins except for convenience. If $A = 10$ and $B = 9$, then that system also works, since you can pay a bill of a single talent by giving one A coin, and taking a B coin in change. *If you can pay 1 talent then you can pay any number, and the system works.*

Euclid's Coin Problem comes down to, given two positive integers A and B, finding integers s and t so that $sA + tB = 1$.

From that equation we see that it does not work to take $A = 25$ and $B = 15$, because then it would only be possible to pay in multiples of their common divisor, 5 talents. We must have $\gcd(A, B) = 1$, the values A and B must be coprime. Is that condition sufficient?

Example 7.11. Let $A = 58$ and $B = 21$. Using the Euclidean Algorithm:

$$
\begin{array}{rcll}
58 & = & 21 \cdot 2 + 16 & \text{(scale by 4)} \\
21 & = & 16 \cdot 1 + 5 & \text{(scale by } -3) \\
16 & = & 5 \cdot 3 + 1 & \text{(scale by 1)} \\
5 & = & 1 \cdot 5 + 0 &
\end{array}
$$

we find that $\gcd(58, 21) = 1$, that is, the coins A and B are coprime, so it is conceivable that there is a solution.

In fact, the four equations above for the divisions with remainder in the Euclidean algorithm hold the key to finding the solution to Euclid's coin problem. Adding the three upper equations, appropriately scaled, will cause the terms in the unwanted remainders, 5 and 16, to cancel out. The surviving terms combine into $(4)(58) + (-3)(21) = (8)(21) + 1$, or $(4)(58) + (-11)(21) = 1$. So with coins $A = 58$ and $B = 21$ we can pay one talent by paying 4 Alexanders, and getting back 11 Bucephaluses in change.

How were the scaling factors chosen? Bottom to top. Factor 1 was chosen so that the $\gcd(58, 21) = 1$ would not be altered. Scaling factor -3 was chosen to cancel out the two terms for remainder 5. Then scaling factor 4 was chosen to cancel out the three terms for remainder 16.   ◇

Euclid's Coin Problem is not just recreational mathematics. It turns out to be an important problem in understanding number theory, and the Euclidean Algorithm has an extension specifically designed to compute the scaling factors needed to solve the coin problem. This is an alternative to the linear algebra approach of the previous example.

Algorithm 7.12 (Extended Euclidean Algorithm). To find $\gcd(n_0, n_1)$ with $n_0 > n_1$; in the kth step, $k \geq 0$, perform division with remainder $n_k = n_{k+1} \cdot q_{k+1} + n_{k+2}$ to define the new value n_{k+2}. The procedure halts at the Kth step when $n_{K+2} = 0$. Then $\gcd(n_0, n_1) = n_{K+1}$.

Now set $s_K = 1$ and $s_{K-1} = -q_K$, and compute $S_{K-2}, \ldots, S_0$ using $s_k = -s_{k+1} \cdot q_{k+1} + s_{k+2}$, halting at s_0. Then $s_0 n_1 + s_1 n_0 = \gcd(n_0, n_1)$.   ♡

Here is the same example done via the extended algorithm,

$$
\begin{aligned}
58 &= 21 \cdot 2 + 16 & s_0 = (-11) &= -(4) \cdot 2 + (-3) \\
21 &= 16 \cdot 1 + 5 & s_1 = (4) &= -(-3) \cdot 1 + (1) \\
16 &= 5 \cdot 3 + 1 & s_2 = (-3) & \quad s_3 = (1); \\
5 &= 1 \cdot 5 + 0
\end{aligned}
$$

So $4 \cdot 58 - 11 \cdot 21 = 1$. The extension is a little bit more work because of the negatives, but there is no division.

To show that the extended algorithm is correct and the scaling factors s_k are chosen correctly, we use a backward induction, with the statement being that, at the kth step of the extension, $s_k n_{k+1} + s_{k+1} n_k = \gcd(n_0, n_1)$.

Proof. For the base case, $k = K - 1$, we have $n_{K-1} = n_K q_K + \gcd(n_0, n_1)$, or $n_{K-1} + (-q_K)n_K = \gcd(n_0, n_1)$, or $s_K n_{K-1} + s_{K-1} n_K = \gcd(n_0, n_1)$, as required.

For the inductive step, let k be given and assume that $s_{k+2} n_{k+1} + s_{k+1} n_{k+2} = \gcd(n_0, n_1)$. We have $s_k = -s_{k+1} \cdot q_{k+1} + s_{k+2}$ and $n_k = n_{k+1} \cdot q_{k+1} + n_{k+2}$, so substituting in s_{k+2} and n_{k+2} gives $(s_k + s_{k+1} q_{k+1})n_{k+1} + s_{k+1}(n_k - n_{k+1} q_{k+1}) = \gcd(n_0, n_1)$. Then canceling $s_{k+1} q_{k+1} n_{k+1}$ gives $s_k n_{k+1} + s_{k+1} n_k = \gcd(n_0, n_1)$, as required. □

Theorem 7.13. *Given positive integers A and B, there exist integers s and t with $sA + tB = n$ if and only if $\gcd(A, B) \mid n$.*

So we not only know when the coin problem has a solution, using the extended Euclidean algorithm, we can find the required numbers "quickly."

Let's now return to the 13^2 problem. Why, just by the primality of 11, 13, and 19 is it impossible that $13^2 = 11 \cdot 19$?

We may not know much more about primality, but we now know a lot more about coprimality. The numbers 13 and 19 are coprime, so we can solve Euclid's Coin Problem: there are numbers s and t with $s \cdot 13 + t \cdot 19 = 1$. Multiplying both sides by 13 gives $s \cdot 13^2 + t \cdot 19 \cdot 13 = 13$. But if $13^2 = 11 \cdot 19$, then we could substitute: $s \cdot 11 \cdot 19 + t \cdot 19 \cdot 13 = 13$, or $(s \cdot 11 + t \cdot 13)19 = 13$, which says $19 \mid 13$, violating 13's primality.

Notice that we did no arithmetic, and we did not even bother to actually figure out the solution to the coin problem. We just used the fact that there *was* a solution.

The argument solving the 13^2 problem, with just a little attention to detail is used in Section 7.9 to show that integers satisfy *unique prime factorization*.

Theorem 7.14. *Every integer n, n > 1 factors uniquely into primes.*

Exercises

1. Suppose Alexander is worth 34 talents and the Bucephalus is worth 21 talents. How can you pay a bill of 7 talents?
2. Find s and t such that $s \cdot 449 + t \cdot 106 = 1$.
3. Find s and t such that $s \cdot 111 + t \cdot 99 = 1$ or prove that none exists.

7.4   Modular arithmetic

When solving mathematical problems and puzzles, it is often the case that an argument or insight will turn on the fact that some particular number is even or odd. This is not simply because there are two classes, the even numbers divisible by 2, and the odd numbers not. There are lots of other ways to split the numbers into two classes. Evenness and oddness is special because integer arithmetic respects it. You have seen this fact written in pseudo-equations like

$$
\begin{array}{rclcrcl}
\text{Even} + \text{Even} & = & \text{Even} & \qquad & \text{Even} \times \text{Even} & = & \text{Even} \\
\text{Odd} + \text{Odd} & = & \text{Even} & & \text{Odd} \times \text{Odd} & = & \text{Odd} \\
\text{Even} + \text{Odd} & = & \text{Odd} & & \text{Even} \times \text{Odd} & = & \text{Even}
\end{array}
$$

Why not try the same split to express threeness, and call those numbers divisible by 3 "threven", and those not "throdd". But that setup does not respect the arithmetic: 2 and 3 are throdd, and $2 + 3 = 5$, which would say that adding two throdds gives a throdd, but 4 and 5 are also both throdd, and $4 + 5 = 9$, threven.

You have probably already anticipated from division with remainder that we need three classes. So let's redefine throdd as those numbers which, when divided by 3, leave a remainder of 1, and invent a new word thweird, for those which leave a remainder of 2. Threven numbers have remainder 0 when divided by 3, so now every number is in exactly one of the classes – threven, throdd, or thwierd. Now we have new pseudo equations and if you test, you find that they all actually do work, like

$$
\begin{array}{rclcrcl}
\text{Threven} + \text{Threven} & = & \text{Threven} & \qquad & \text{Threven} \times \text{Threven} & = & \text{Threven} \\
\text{Throdd} + \text{Throdd} & = & \text{Thwierd} & & \text{Throdd} \times \text{Throdd} & = & \text{Throdd} \\
\text{Throdd} + \text{Thweird} & = & \text{Threven} & & \text{Throdd} \times \text{Thweird} & = & \text{Thweird}
\end{array}
$$

and there are several more. Everything favors pursuing this analogy in the study of the trinary aspects of the natural world except the notation. Not only are the words ugly and awkward, we will need many more of them, since we will also want to pursue the fourfold analogy, and the fivefold . . . .

So let's proceed immediately to the general case, to the n-fold analogy to even/odd in which the number n is called the *modulus*. Two integers a and b are said to be *congruent modulo n*, and we write $a \equiv b$ mod n, if $a = b + kn$ for some $k \in \mathbb{Z}$. Congruence modulo n is a relation which is reflexive, symmetric, and transitive; hence congruence modulo n is an equivalence relation. We are particularly interested in the equivalence classes since they are exactly the sets of numbers which have the same remainder when divided by n. There are n equivalence classes, one for each remainder $0, 1, 2, \ldots (n-1)$, and *those symbols are used to denote the equivalence classes*. The set of equivalence classes for modulus n is denoted by $\mathbb{Z}_n = \{0, 1, 2, \ldots, n-1\}$.

For modulus 2, the equivalences classes are the even and odd numbers, with the even numbers denoted by 0 and the odd numbers denoted by 1, $\mathbb{Z}_2 = \{0, 1\}$.

For modulus 3, $\mathbb{Z}_3 = \{0, 1, 2\}$, and the symbol 1 stands for the equivalence class of numbers whimsically called throdd above. This *abuse of notation* actually causes less confusion that you might imagine, but you should still be cautious. We know that the number 1 is not equal to the set of odd numbers, and the set of odd numbers is not equal to the set of numbers which have remainder 1 when divided by 3; but we are now using the symbol "1" for all three of them, and many others.

But the set theory is secondary. What gives the numbers modulo n so much power is the arithmetic. The classes with the same modulus can be added and multiplied consistently, just as the even and odd numbers can.

Theorem 7.15. *If $a \equiv a'$ mod n and $b \equiv b'$ mod n then $a + b \equiv a' + b'$ mod n and $ab \equiv a'b'$ mod n.*

Proof. We have $a = a' + kn$ and $b = b' + jn$ for some integers k and j. Adding gives $a + b = a' + kn + b' + jn = (a' + b') + (k + j)n$, which implies $a + b \equiv a' + b'$ mod n. Multiplying gives $ab = (a' + kn)(b' + jn) = a'b' + b'kn + a'jn + kjn^2 = a'b' + (b'k + a'j + kjn)n$, so $ab \equiv a'b'$ mod n. $\square$

Even better, this *modular arithmetic* inherits all the important properties of arithmetic in $\mathbb{Z}$. Addition and multiplication in $\mathbb{Z}_n$ is commutative and associative, and the distributive law holds. There is an additive unit, always written with symbol 0, and a multiplicative unit, with symbol 1. You will have to practice this new arithmetic, in which $7 + 6 \equiv 13$ mod 20, $7 + 6 \equiv 0$ mod 13, and $7 + 6 \equiv 4$ mod 9. Each set $\mathbb{Z}_n$ has its own addition and multiplication table, such as the ones below for $\mathbb{Z}_5$.

+	0	1	2	3	4
0	0	1	2	3	4
1	1	2	3	4	0
2	2	3	4	0	1
3	3	4	0	1	2
4	4	0	1	2	3

×	0	1	2	3	4
0	0	0	0	0	0
1	0	1	2	3	4
2	0	2	4	1	3
3	0	3	1	4	2
4	0	4	3	2	1

Note that, if the modulus is clear from the context, many authors will abbreviate $17 + 6 \equiv 10 \bmod 13$ to $17 + 6 \equiv 10$, or often just $17 + 6 = 10$.

Exercises

1. For each of the following, compute the results modulo 2, 3, 5, 10, and 11. In each case express the answer from 0 to the modulus minus one: $6 + 7$, $8 + 8$ $2 \cdot 8$, $1 + 2 + 3 + 4 + 5 + 6 + 7 + 8 + 9$, $5280 + (65)(88)$, $5280 - (65)(88)$.
2. Make addition and multiplication tables modulo 7. Find all pairs of additive and multiplicative inverses.
3. Make addition and multiplication tables modulo 8.
   Find all pairs of additive and multiplicative inverses.

7.5 Multiplicative inverses

In $\mathbb{Z}$, every element k has an additive inverse $-k$, since $k + (-k) = 0$. In $\mathbb{Z}_n$ every element k has the additive inverse $n - k$, since $k + (n - k) = n \equiv 0 \bmod n$.

In $\mathbb{Z}$, only 1 and -1 have multiplicative inverses, and that is only because $1^2 = (-1)^2 = 1$. Otherwise, if you want to divide, you have to move to the rational numbers where division is always possible except by 0.

In $\mathbb{Z}_n$, the situation is in between that of $\mathbb{Z}$ and $\mathbb{Q}$. If i and j are multiplicative inverses modulo n, then $ij \equiv 1 \bmod n$, or there exists a $k \in \mathbb{Z}$ so that $ij = 1 + kn$, or $ij + (-k)n = 1$. So, given n and i, we want to find integers j and k so that $ij + (-k)n = 1$. But that is exactly Euclid's Coin Problem, and we already know that there is a solution exactly when $\gcd(i, n) = 1$, and then we can use the Extended Euclidean Algorithm to find it.

Theorem 7.16. *The number k has a multiplicative inverse in $\mathbb{Z}_n$ if and only if* $\gcd(k, n) = 1$.

Example 7.17. Find the multiplicative inverse of 1776 modulo 2021. Find a few other pairs of multiplicative inverses in $\mathbb{Z}_{2021}$.

✠

For 1776, we use the Extended Euclidean Algorithm:

$$
\begin{aligned}
2021 &= 1776 \cdot 1 + 245 & (-33) &= -(29) \cdot 1 + (-4) \\
1776 &= 245 \cdot 7 + 61 & (29) &= -(-4) \cdot 7 + (1) \\
245 &= 61 \cdot 4 + 1 & s_2 = (-4); & \quad s_3 = (1)
\end{aligned}
$$

So $(29)(2021) + (-33)(1776) = 1$, which says $(-33)(1776) \equiv 1 \bmod 2021$, or the multiplicative inverse of 1776 in $\mathbb{Z}_{2021}$ is $2021 - 33 = 1988$.

To find some other pairs, we could pick some numbers at random, and follow the same procedure, but consider that we want to "factor 1", and $1 \equiv 2022 \bmod 2021$. Factoring $2022 = 2 \cdot 1011 = 2 \cdot 3 \cdot 337$. So we have multiplicative inverses $(2)(1011) = (3)(674) = (6)(337)$. Another trick is to note that $-1 \equiv 2020 \bmod 2021$, also easy to factor: $2020 = 202 \cdot 10 = 2^2 \cdot 5 \cdot 101$. That gives us multiplicative inverses $(4)(-505) = (-5)(505) = (20)(-101) = (101)(-20)$, or $(4)(1516)$, $(2016)(505)$, $(20)(1920)$, and $(101)(2001)$, respectively.

Some elements in $\mathbb{Z}_{2021}$ have no multiplicative inverse because 2021 is not prime, having prime factorization $2021 = 43 \cdot 47$. So no multiple of 43 or 47 can have a multiplicative inverse modulo 2021. $\diamond$

Multiplicative inverses are the key to division in $\mathbb{Z}_n$, just as in $\mathbb{Q}$ where $i/j = i \cdot j^{-1}$. Another way of looking at it, without the negative exponents, is that dividing, $2/3 = x$, is the same as solving the equation $3 \cdot x = 2$ for x, which you do by multiplying both sides by the reciprocal of 3, that is, by its multiplicative inverse.

Example 7.18. Compute $2/3$ in $\mathbb{Z}_{11}$ and in $\mathbb{Z}_{12}$. Equivalently, solve $3x = 2$ in $\mathbb{Z}_{11}$ and $\mathbb{Z}_{12}$.

✠

For $\mathbb{Z}_{11}$, we need the multiplicative inverse of 3. We can always use the Extended Euclidean Algorithm, but it is never a bad idea to look for a quick shortcut first. $1 \equiv 12 \bmod 11$, and $12 = 3 \cdot 4$, so the multiplicative inverse of 3 in $\mathbb{Z}_{11}$ is 4. So $3x = 2$ gives $4 \cdot 3x = 4 \cdot 2$. Modulo eleven we have

$$(4 \cdot 3)x \equiv x \equiv 2 \cdot 4 \equiv 8 \bmod 11 \text{ hence } 2/3 \equiv 8 \bmod 11.$$

In $\mathbb{Z}_{12}$, the number 3 has no multiplicative inverse. But maybe we can solve $3x \equiv 2 \bmod 12$ anyway. That would mean $3x = 2 + k \cdot 12$ for some k. That would give us $3x - 12k = 2$ or $3(x - 4k) = 2$, which is impossible since 3 does not divide 2. We conclude that $2/3$ cannot be defined in $\mathbb{Z}_{12}$, and the equation $3x \equiv 2 \bmod 12$ has no solution. (The obstruction was that 2 was not divisible by 3. What about the solving $3x \equiv 9 \bmod 12$?) $\diamond$

Exercises

1. Find the multiplicative inverse of 5 modulo 66.
   Solve the equation $5x = 3$ in $\mathbb{Z}_{66}$.
2. Find the multiplicative inverse of 12 modulo 25.
   Compute $5/12$ modulo 25.
3. Find all multiplicative inverses in $\mathbb{Z}_{25}$.

7.6 The Chinese Remainder Theorem

Suppose that on Chinese New Year there was a marching band on parade. On Main Street the band was marching eight abreast, but on Park Avenue the band had to reorganize and was marching eleven abreast. What can you say about the number of musicians in the band?

The number of musicians is a multiple of 8 and of 11, so $n = 2^3 k = 11 j$. By unique prime factorization we conclude n is divisible by $2^3 11 = 88$.

Theorem 7.19. *If a and b are coprime and $a \mid n$ and $b \mid n$, then $ab \mid n$.*

Proof. We have $n = ai = bj$. Create a prime factorization of n by multiplying prime factorizations of a and i. Form a second one by multiplying prime factorizations of b and j. By unique prime factorization, Theorem 7.14, the prime factorizations of n are the same up to reordering. Since $\gcd(a, b) = 1$, none of the primes for a occur in the factorization of b, so the primes in the prime factorization for a must all be in the prime factorization for j, that is $a \mid j$, that is $j = ak$ for some $k \in \mathbb{Z}$, hence $n = b(ak) = (ab)k$. □

Let's return to the parade problem. Suppose that, on more careful examination, it is found that on Main Street the last row was short, having only three men, while on Park Avenue the last row was also short, having only two men. Would those few men change anything?

The number theoretic equations change. From the information on Main Street we have $n \equiv 3 \bmod 8$ and from Park Avenue we know $n \equiv 2 \bmod 11$.

One way to proceed is to check one by one all the integers consistent with Main Street, $3 + k \cdot 8$, so $3, 11, 19, 27, 35, \ldots$, against the requirement of Park Avenue. This is sufficient for a small problem like this, but will not scale.

It is more effective to take the general expression $3 + k \cdot 8$ for the solutions for Main Street and to solve $3 + k \cdot 8 \equiv 2 \bmod 11$ for k to find which also work for Park Avenue. We quickly get $k \cdot 8 \equiv -1 \bmod 11$, and the next step to isolate

k is to multiply by the multiplicative inverse of 8 modulo 11:

$$
\begin{aligned}
11 &= 8 \cdot 1 + 3 & -4 &= -(3) \cdot 1 + (-1) \\
8 &= 3 \cdot 2 + 2 & 3 &= -(-1) \cdot 2 + (1) \\
3 &= 2 \cdot 1 + 1 & s_2 = -1 & \qquad s_3 = 1; \\
2 &= 1 \cdot 2 + 0
\end{aligned}
$$

So $3 \cdot 11 - 4 \cdot 8 = 1$, and the multiplicative inverse of 8 is -4. Now we can solve for k by $k \cdot 8(-4) \equiv (-1)(-4) \equiv 4 \bmod 11$, so $k = 4$. We find that the number of musicians in the parade could be $3 + 4 \cdot 8 = 35$. That is less than half the minimum computed for the first version of the problem, so obviously those few stray men at the end do make a difference.

It seems amazing that you can figure out the number of musicians from such scanty information, but in fact we did no such thing. We just found the minimum possible. Any multiple of 88 can be added to get another possible population for the band: $35 + j \cdot 88$. So maybe the band had 123 members, but no number in between 35 and 123 is possible.

Theorem 7.20 (Chinese Remainder Theorem). *Given two simultaneous congruences: $x \equiv a \bmod n$ and $x \equiv b \bmod m$ with $\gcd(n, m) = 1$. If $s \cdot n + t \cdot m = 1$ for some $s, t \in \mathbb{Z}$, then the solutions are*

$$
bs \cdot n + at \cdot m + knm
$$

for $k \in \mathbb{Z}$.

Proof. The equation guaranteed by the Extended Euclidean Algorithm $s \cdot n + t \cdot m = 1$ implies that $bs \cdot n \equiv b \bmod m$ and $bs \cdot n \equiv 0 \bmod n$; while $at \cdot m \equiv 0 \bmod m$ and $at \cdot m \equiv a \bmod n$. Adding them $x = bs \cdot n + at \cdot m$ gives the desired result modulo n and modulo m.

So x is one solution. If y is another, then $x - y \equiv a - a \equiv 0 \bmod n$, so $n \mid x - y$, and also $m \mid x - y$. Since $\gcd(n, m) = 1$, $nm \mid x - y$, and $y = bs \cdot n + at \cdot m + knm$. $\qquad \square$

What if the two moduli are not coprime? There may be a solution: $n \equiv 11 \bmod 45$ and $n \equiv 38 \bmod 63$ has solution $n = 101$. On the other hand, there may not: $n \equiv 16 \bmod 45$ and $n \equiv 32 \bmod 63$ has no solution whatsoever.

This aspect of the problem was also worked out hundreds of years ago and with the advantage of our modern notation, you have a pretty good chance of figuring it out.

Exercises

1. Find any integer n such that $n \equiv 17 \bmod 19$ and $n \equiv 11 \bmod 13$.
2. Find all integers n such that $n \equiv 2 \bmod 19$ and $n \equiv 5 \bmod 13$.
3. Find any integer n such that $3n \equiv 17 \bmod 19$ and $5n \equiv 11 \bmod 13$.

7.7 Case study: Diophantus

Diophantus was a mathematician living in the 3rd century AD in Alexandria, when that Egyptian city had long been Greek, although part of the Roman Empire, for over a century. Algebraic equations for which integer solutions are required, such as $a^2 + b^2 = c^2$ with solution $3^2 + 4^2 = 5^2$, were his speciality and are still called *Diophantine* after him.

Long after Diophantus passed away, the writer Metrodoros heard the following number theoretic puzzle, and included it into his collection of epigrams, about one and a half millennia ago. This is Paton's close translation from Greek.

Diophantus

This tomb holds Diophantus. Ah, how great a marvel!
The tomb tells scientifically the measure of his life.
God granted him to be a boy for the sixth part of his life,
    and adding a twelfth part to this,
He clothed his cheeks with down;

He lit him the light of wedlock after a seventh part,
    and five years after his marriage he granted him a son.
Alas! Late-born wretched child;
after attaining the measure of half of his father's full life,
    chill Fate took him.

After consoling his grief by this science of numbers for four years
    he ended his life.

Metrodoros, Epigram 126

Metrodoros gave his epigrams in Greek verse, so you might prefer this more modern poetical rendering without the strange Greek idioms:

"Here lies Diophantus," the wonder behold.
Through art algebraic, the stone tells how old:
"God gave him his boyhood one-sixth of his life,
One twelfth more as youth while whiskers grew rife;
And then yet one-seventh ere marriage begun;
In five years there came a bouncing new son.

Alas, the dear child of master and sage
Met fate at just half of his dad's final age
Four years yet his studies gave solace from grief:
Then leaving scenes earthly he, too, found relief."

which was old when reprinted in the "Yearbook" of the National Council of Teachers of Mathematics, 1926.

The challenge is to work out the chronology of the life of Diophantus, ideally without resorting to a computer or even the use of pencil and paper.

✠

This first key to the puzzle is to understand that the solution is to be a whole number of years, and approach it as a Diophantine problem, not as one in which we want to measure his life on a continuous time scale. The second is to notice that in one couplet his age in years is declared divisible by 12, and another to be divisible by 7. Since $\gcd(7, 12) = 1$, unique prime factorization requires his age to be divisible by $7 \cdot 12 = 84$. Unless Diophantus were truly marvelously old, that gives his age as 84, and the remaining spans of years are then easy to unravel. No equations and algebra, just number theory.

One can also use number theory on the Guarini Problem, the case study of Section 1.9, a problem in recreational mathematics inspired by chess. The problem can be successfully recast and attacked using $\mathbb{Z}_9$. Try it.

Consider also the next puzzle which Sam Loyd syndicated in the newspapers of his day, about the turn of the previous century. Loyd asked "How can you score exactly 50 points?"

Modular arithmetic holds the key both to how the puzzle was designed, the way to a super fast solution, and the fact that the solution is unique. Pedestrian puzzlists trying methodically each selection of 3 or 4 dolls and adding their tags will be looking at $\binom{10}{3} + \binom{10}{4} = 120 + 210 = 330$ tasks.

✠

Looking at the puzzle modulo 2, the sum of the dolls must be congruent to 0, and you conclude that you need an even number of odd dolls – but that is not the best modulus...

Ready for more, or want to design your own? New and original puzzles involving number theory appear all the time. The Pi Mu Epsilon Journal, for example, always has a section for new and original problems accessible to students.

7.8 Case study: The Indian formulas

Most people know that the lengths of the sides of a right triangle follow Pythagoras's Theorem, $c^2 = a^2 + b^2$, or $c = \sqrt{a^2 + b^2}$. Because of the square root, the length of the hypotenuse is expected to be irrational even if the leg-lengths are integers, like for a triangle with sides 1, 1, and $\sqrt{2}$. Since $3^2 + 4^2 = 5^2$, however, there is at least one right triangle, the 3-4-5 triangle, for which all three sides are integers. Are there any others? Yes. They have been known for a long time to those who study number theory and are called *Pythagorean Triples*.

The method in this case study is to start with the assumption that we have a Pythagorean Triple, work out what properties the numbers in the triple must have, and eventually derive a formula for them. If some of the steps seem haphazard or arbitrary, it is probably because they are very clever shortcuts, worked out over a very long time, from a much longer journey.

Suppose a, b, and c are all natural numbers, and suppose $c^2 = a^2 + b^2$. Any prime which divides two of them, divides the third and simply expands the size of the triangle by an integer factor, which is not very interesting. So let's assume that $\gcd(a, b) = 1$, and that means $\gcd(a, c) = \gcd(b, c) = 1$ as well.

Consider the equation $c^2 = a^2 + b^2$ modulo 4. (Shortcut!) Modulo 4, odd numbers have square 1, since $(2k + 1)^2 = 4k^2 + 4k + 1 \equiv 1 \bmod 4$, and even numbers have square 0, since $(2k)^2 = 4k^2 \equiv 0 \bmod 4$. That means that a and b cannot be both odd, since then $c^2 \equiv 1 + 1 \equiv 2 \bmod 4$, which is not possible for a square. To avoid duplicates, assume from now on that b is the even leg of the triangle, and since $\gcd(a, b) = 1$, leg a must be odd, and so the hypotenuse c must also be odd.

Write $b^2 = c^2 - a^2 = (c + a)(c - a)$. Since a and c are both odd, $c + a$ and $c - a$ are both even. So $(c + a)/2$ and $(c - a)/2$ are both integers.

The integers $(c + a)/2$ and $(c - a)/2$ must be coprime. That is because any prime dividing both of them would divide their sum and their difference. But $(c + a)/2 + (c - a)/2 = c$ and $(c + a)/2 - (c - a)/2 = a$. And we know $\gcd(a, c) = 1$.

The integer $b^2/4$ is a perfect square, so its prime factors come in pairs. By unique prime factorization, those primes make up the prime factors of $(c + a)/2$ and $(c - a)/2$, since $b^2/4 = [(c + a)/2][(c - a)/2]$. But $(c + a)/2$ and $(c - a)/2$ are coprime, so each also has prime factors grouped in pairs, so each one must also be a perfect square.

Now write $(c + a)/2 = n^2$ and $(c - a)/2 = m^2$ and since $b^2/4 = n^2 m^2$, $b = 2nm$. Now, reusing a trick from above, $c = (c + a)/2 + (c - a)/2 = n^2 + m^2$ and $a = (c + a)/2 - (c - a)/2 = n^2 - m^2$.

So every Pythagorean Triple is obtained from two natural numbers n and m, using the *Indian formulas*.

$$a = n^2 - m^2, \qquad b = 2nm, \qquad c = n^2 + m^2. \tag{7.1}$$

Does every n and m work? Look!

$$(n^2 - m^2)^2 + (2nm)^2 = (n^4 - 2n^2m^2 + m^4) + 4n^2m^2$$
$$= n^4 + 2n^2m^2 + m^4 = (n^2 + m^2)^2$$

so any choice of n and m yield by Eq. (7.1) numbers a, b, and c with $a^2 + b^2 = c^2$. And we just saw that all Pythagorean Triples come from such an n and m.

So is there anything left? Yes. We wanted specifically those Pythagorean Triples with $\gcd(a, b) = 1$. Not all n and m do that. We'll leave it as a puzzle for the reader to show that for this last bit we want $\gcd(n, m) = 1$, and also that n or m should be even.

Anyway, there are lots of triples! So let's make some. If $n = 2$ and $m = 1$, that gives the 3–4–5 triangle. If $n = 4$ and $m = 1$, then the triangle is 15–8–17. If $n = 5$ and $m = 4$, then the triangle is 9–40–41. If $n = 5$ and $m = 2$, then the triangle is 21–20–29.

The last one is interesting, since it is very close to a right triangle with angle 45°. It turns out that there are so many Pythagorean triples that for any acute angle, there is a Pythagorean Triple approximating that angle to any desired accuracy. But that would be yet another case study.

Since there are infinitely many solutions of the Diophantine equation $c^2 = a^2 + b^2$, it is natural to wonder about integer solutions to $c^3 = a^3 + b^3$, or $c^4 = a^4 + b^4$, or .... Can we solve those? People wondered about that even though there is no obvious application, as there is with Pythagorean Triples. But there aren't any.

Theorem 7.21 (Fermat's Last Theorem). *If $n > 2$, then the equation $c^n = a^n + b^n$ has no positive integer solutions.*

7.9 Case study: Unique prime factorization

The solution to the 13^2 problem illustrated the main reason why integers are uniquely factorable into primes. In this case study, we want to examine this issue in more detail. The following theorem contains the heart of the matter.

Theorem 7.22. *Let $p \in \mathbb{P}$ and $p > 1$.*

$$\forall n, m \in \mathbb{N}; \, [p = nm] \Rightarrow [(n = 1) \vee (m = 1)]$$

$$\Updownarrow$$

$$\forall n, m \in \mathbb{N}; \, [p \mid nm] \Rightarrow [(p \mid n) \vee (p \mid m)]$$

Reading through the logical notation, you might recognize that the upper condition simply expresses the definition of the primality of p, that is, that p only factors as p times 1. The second condition, is just what we needed for the 13^2 problem.

The theorem says that the two conditions are equivalent. That equivalence implies that the lower condition could also function as the definition of primality. Actually, in more algebraic expositions, that is exactly what is done. So the lower condition may be regarded as the definition of primality, and the upper condition the definition of being *irreducible*, and the theorem would then be stating that, for $\mathbb{Z}$, primes and irreducibles are the same.

Of course, the proof will use the double implication method.

Proof. Suppose the lower condition is true. To prove the upper one, let n, and m be given, and assume that $p = nm$. So $p = 1 \cdot nm$ and $p \mid nm$. From the lower condition we know that $p \mid n$ or $p \mid m$. If $p \mid n$, then $n = kp$ for some $k \in \mathbb{Z}$, and $p = kpm$. Canceling p gives $km = 1$. Thus k and m are both 1 and $(n = 1) \vee (m = 1)$, as required. If $p \mid m$, exactly the same argument applies. So $[p = nm] \Rightarrow [(n = 1) \vee (m = 1)]$, completing the first half. (That was the easy half.)

Suppose now that the upper condition is true. To prove the lower one, let n and m be given, and assume that $p \mid nm$, that is $pk = nm$ for some integer k.

Case 1: $p \mid n$. Then $(p \mid n) \vee (p \mid m)$.

Case 2: $\neg(p \mid n)$. Then $\gcd(p, n) = 1$. So (Euclid's Coin Problem again!) there are integers s and t with $sp + tn = 1$. Multiplying by m, we have $spm + tnm = m$, and substituting $pk = nm$ gives $spm + tpk = m$, or $p(sm + tk) = m$. So $p \mid m$, hence $(p \mid n) \vee (p \mid m)$.

So $[p \mid nm] \Rightarrow [(p \mid n) \vee (p \mid m)]$, concluding the second part. $\square$

Now we can show unique prime factorization! We just had a fancy exercise in the double inclusion method and implications. The next proof is a fancy exercise in induction.

Theorem 7.23. *Let* $p_1 p_2 \cdots p_k = p'_1 p'_2 \cdots p'_{k'}$ *be an equation with all prime factors. Then* $k = k'$, *and the prime factors are the same except for permuting their order.*

Proof. The proof will be by induction on $k + k'$ and we have to show the proposition is true for all $k + k' \geq 2$.

Base case: $k + k' = 2$. Since neither side can be 1, each side contains a prime, so $k = k' = 1$, and $p_1 = p'_1$.

Inductive step. Let $k + k'$ be given and assume the strong induction hypothesis, that is, unique factorization holds true for all shorter prime expressions. We have $p_1 \mid p'_1 p'_2 \cdots p'_{k'}$.

Let's prove by induction on k' that p_1 is one of the primes in the product $p'_1 p'_2 \cdots p'_{k'}$ for all $k' \geq 1$. If $k' = 1$, there is nothing to show, establishing the base case. Now $p_1 \mid p'_1 p'_2 \cdots p'_{k'} = (p'_1)(p'_2 \cdots p'_{k'})$, so by Theorem 7.22 $p_1 \mid p'_1$ or $p_1 \mid (p'_2 \cdots p'_{k'})$. If the first is true, then $p_1 = p'_1$. If the second is true, then p_1 is one of the primes in the product $p'_2 \cdots p'_{k'}$ by the induction hypothesis. So in either case, p_1 is one of the primes on the list.

Now, to continue with the main induction, $p_1 = p'_i$ for some i. So we can cancel those two primes and get a shorter expression which, by the (main) inductive hypothesis is just the same primes up to permuting the factors. Thus the same is true replacing the factors $p_1 = p'_i$. □

One induction inside the induction step of another induction! What could be fancier than that?

7.10 Summary exercises

You should have learned about:

- The definition of prime number.
- The sieve method for finding primes.
- The use of the Euclidean Algorithm to compute $\gcd(n, m)$, and its efficiency.
- The use of the Extended Euclidean Algorithm to compute s and t so that $sn + tm = \gcd(n, m)$.
- Basic modular arithmetic.
- How to compute multiplicative inverses modulo n, that is, in $\mathbb{Z}_n$.
- That for p a prime, division makes sense in $\mathbb{Z}_p$.
- How to solve equations of the type $ax \equiv b \bmod n$.
- How to use the Chinese Remainder Theorem to solve pairs $ax \equiv b \bmod n$, $cx \equiv d \bmod m$ of congruences if $\gcd(n, m) = 1$.

1. Let p and q be primes. Show that $p^2 - q^2$ is not prime.
2. How many integers are there that are not divisible by any prime larger than 64 and not divisible by the cube of any prime?
   Show all your work.
3. Show that $\gcd(n, m) = \gcd(n - m, m)$. Give an example to show that $\gcd(n, m) = \gcd(n - m, n + m)$ is not always true. Can it ever be true?
4. Let $n \geq 5$. What can you say about $k = \gcd(n + 2, n - 2)$? How many different values can k have? Give an example of each type.
5. Suppose a country has coins in denominations 5, 15, and 27 Tzarlinkas. Can you find a way to use them to pay a 101 Tzarlinka bill?
6. Let p, q, and r be three distinct primes. Show that $spq^2 + tqp^2 = r$ has no solutions s and t in the integers.
7. Show that the multiplicative inverse of $n - 1$ is always $n - 1$ modulo n.
8. A band of trombones tries to march in rows of 8 but the last row has only 4 members. It is reorganized to march in rows of 11 but then there are only 10 members in the final row. What is the fewest number of members this band can have?
9. Find a number m such that $m \equiv 3 \bmod 5$, $m \equiv 5 \bmod 7$, and $m \equiv 7 \bmod 11$.

10. Let k be given. Find two numbers n and m, $n > m$, such that $n \equiv m \bmod 3^k$ but $n \not\equiv m \bmod 3^{k+1}$.

11. Many people know the equation $3^2 + 4^2 = 5^2$. Show that every Pythagorean triple, $a^2 + b^2 = c^2$ with $a, b, c \in \mathbb{N}$, has one side divisible by 3, one side divisible by 4, and one side divisible by 5.

12. Let N be a fixed natural number. Define a relation $\equiv$ on the integers by setting $n \equiv m \pmod{N}$ if $n - m$ is evenly divisible by N.
Show that this relation is an equivalence relation.

Chapter 8

Codes and cyphers

8.1 Exponentials modulo n

Suppose we wanted to compute 8^5 mod 3. We can try to simplify the exponent, take $5 \equiv 2$ mod 3 and use $8^2 = 64 \equiv 1$ mod 3. That is certainly easier, but wait. Suppose we follow a different path, note that the base satisfies $8 \equiv 2$ mod 3 and compute $8^5 \equiv 2^5 = 32 \equiv 2$ mod 3. There seems to be a discrepancy. Do we have a contradiction to Theorem 7.15? Is the answer 1 or 2?

It is 2. The first path was invalid. Theorem 7.15 says that modular equivalence respects addition and multiplication, but says nothing about exponents. The 'discrepancy' above illustrates that we must not assume that a^i and a^j are equivalent modulo n just because the exponents satisfy $i \equiv j$ mod n. It is the bases of an exponential which behave as expected, as in the next example.

Example 8.1 (The Rule of Nine). The remainder of a number modulo 9 is same as the remainder of the sum of its digits.

Let n be a k digit decimal, $n = d_{k-1} \cdots d_0$, so we may write $n = \sum_{i=0}^{k-1} d_i 10^i$. Since $10 \equiv 1$ mod 9, we have $10^i \equiv 1^i \equiv 1$ mod 9. So $n \equiv d_0 + d_1 + \cdots d_{k-1}$ mod 9.

So 10607 has the same remainder modulo 9 as $1 + 6 + 7 = 14$, or 5. ◇

What can we say about a^j focusing on the exponents as independent variables? Here is a table of modulo 10, which is easy to check since multiplication modulo 10 in base 10 simply ignores all but the 1's digit:

$\mathbb{Z}_{10}$	a^1	a^2	a^3	a^4	a^5	a^6	a^7	a^8	a^9	a^{10}	a^{11}	a^{12}	$\cdots$
0	0	0	0	0	0	0	0	0	0	0	0	0	
1	1	1	1	1	1	1	1	1	1	1	1	1	
2	2	4	8	6	2	4	8	6	2	4	8	6	
3	3	9	7	1	3	9	7	1	3	9	7	1	
4	4	6	4	6	4	6	4	6	4	6	4	6	
5	5	5	5	5	5	5	5	5	5	5	5	5	
6	6	6	6	6	6	6	6	6	6	6	6	6	
7	7	9	3	1	7	9	3	1	7	9	3	1	
8	8	4	2	6	8	4	2	6	8	4	2	6	
9	9	1	9	1	9	1	9	1	9	1	9	1	

You see various behaviors for the different elements $a \in \mathbb{Z}_{10}$, but one thing is common, as one keeps multiplying by a, sooner or later a duplicate appears, establishing a pattern, and that pattern must continue forever.

So if we want to compute 888^{888} in $\mathbb{Z}_{10}$, we reduce the base, not the exponent, and compute instead 8^{888}. Then, since the powers of 8 repeat every four, we use the fact that $888 \equiv 4 \bmod 4$, and use $8^{888} = 8^4 \equiv 6 \bmod 10$. Of course $666^{666} \equiv 6 \bmod 10$, since the powers of 6 never seem to get off the ground modulo 10.

Notice in the table above that the exponentials modulo 10 all return to their starting values in the a^9 column. It would be nice if that type of thing happened generally, but exponentials can be stranger still. Here is the table for $\mathbb{Z}_8$.

$\mathbb{Z}_8$	a^1	a^2	a^3	a^4	a^5	a^6	a^7	a^8	$\cdots$
0	0	0	0	0	0	0	0	0	$\cdots$
1	1	1	1	1	1	1	1	1	$\cdots$
2	2	4	0	0	0	0	0	0	$\cdots$
3	3	1	3	1	3	1	3	1	$\cdots$
4	4	0	0	0	0	0	0	0	$\cdots$
5	5	1	5	1	5	1	5	1	$\cdots$
6	6	4	0	0	0	0	0	0	$\cdots$
7	7	1	7	1	7	1	7	1	$\cdots$

in which we see the powers of 6 just drop to 0 and where, of course, they must stay forever, and the sequence never returns to 6.

You see more regular behavior if the base a is coprime to the modulus, so the base has a multiplicative inverse modulo n. For such a base, we see in both $\mathbb{Z}_{10}$ and $\mathbb{Z}_8$ that the sequence of powers passes through a sequence of distinct values until it reaches 1, and then the sequence repeats forever. This is always true.

Theorem 8.2. *Let $n \geq 2$ be an integer and let $a \in \mathbb{Z}_n$. Then a has a multiplicative inverse if and only if there is a number $k < n$ so that $a^k \equiv 1$, and $a^i \not\equiv a^j$ for $1 \leq i < j \leq k$.*

The proof is by double implication.

Proof. Suppose first that a has a multiplicative inverse b. If $a = 1$ then take $k = 1$, so assume $a \neq 1$.

We first show that 1 does occur in the list of powers of a. $\mathbb{Z}_n$ has finite cardinality, so the list $a^1, a^2, \ldots$ eventually must have a duplicate, say $a^j \equiv a^i$ with $j > i$. Let b be the multiplicative inverse of a and multiply both sides by b^i, giving $a^j b^i \equiv a^i b^i$, or $a^{j-i} \equiv 1$ for an exponent $j - i$ smaller than the duplicate a^j. So take k to be the smallest positive value for which $a^k \equiv 1$.

For the other direction, suppose $a^k \equiv 1$. If $k = 1$, then $a = 1$, and a is its own multiplicative inverse. Otherwise $k > 1$ and $aa^{k-1} \equiv 1$, so a^{k-1} is the multiplicative inverse. $\square$

The number k is called the *multiplicative order* of a in $\mathbb{Z}_n$. The multiplicative order is only defined for elements with multiplicative inverses. Also, if a has a multiplicative inverse, we write its multiplicative inverse as a^{-1}, and the usual rules of negative exponents apply, including $a^0 = 1$.

Exercises

1. Compute 10^{200} modulo 11.
2. a) Compute the first 12 powers of 2 modulo 12. Use what you discover to compute 2^{1776} modulo 12.
   b) Compute the first 17 powers of 2 modulo 17. Use what you discover to compute 2^{1776} modulo 17.
3. Compute 2^{1776} modulo 33.

8.2 Prime modulus

When people want to define a modular system for some application, say the 24 hours on a clock, or the 60 seconds on a clock, or the 360 angular degrees around a circle, they seem always to prefer a modulus with many divisors. Whenever such a choice is made, you can be pretty sure that the system is being used primarily additively, that multiplication plays a lesser role. If multiplication is important, then so is division, and every divisor of the modulus loses multiplicative inverses, and that makes the work of solving equations more difficult, or ambiguous, if not impossible.

If you are choosing a modulus, and multiplication is important, you want as few divisors of the modulus as possible. Ideally, you should choose the modulus to be a prime p. In $\mathbb{Z}_p$, every non-zero element has a multiplicative inverse, and so the arithmetic is just the same as it is in $\mathbb{Q}$ or $\mathbb{R}$.

Even the tricky exponents are better behaved if the modulus is prime. In $\mathbb{Z}_{11}$, since every non-zero element has a multiplicative inverse, we have from Theorem 8.2 that the list of powers of every non-zero element will have a 1 before a^{11}. Check that in the table below.

a	a^0	a^1	a^2	a^3	a^4	a^5	a^6	a^7	a^8	a^9	a^{10}	a^{11}
0		0	0	0	0	0	0	0	0	0	0	0
1	1	1	1	1	1	1	1	1	1	1	1	1
2	1	2	4	8	5	10	9	7	3	6	1	2
3	1	3	9	5	4	1	3	9	5	4	1	3
4	1	4	5	9	3	1	4	5	9	3	1	4
5	1	5	3	4	9	1	5	3	4	9	1	5
6	1	6	3	7	9	10	5	8	4	2	1	6
7	1	7	5	2	3	10	4	6	9	8	1	7
8	1	8	9	6	4	10	3	2	5	7	1	8
9	1	9	4	3	5	1	9	4	3	5	1	9
10	1	10	1	10	1	10	1	10	1	10	1	10

One thing which is striking about this table is the column of 1's precisely at a^{10}. We see the same phenomenon in the table of exponents modulo the prime 13.

a	a^0	a^1	a^2	a^3	a^4	a^5	a^6	a^7	a^8	a^9	a^{10}	a^{11}	a^{12}	$\ldots$
0		0	0	0	0	0	0	0	0	0	0	0	0	$\ldots$
1	1	1	1	1	1	1	1	1	1	1	1	1	1	
2	1	2	4	8	3	6	12	11	9	5	10	7	1	
3	1	3	9	1	3	9	1	3	9	1	3	9	1	
4	1	4	3	12	9	10	1	4	3	12	9	10	1	
5	1	5	12	8	1	5	12	8	1	5	12	8	1	
6	1	6	10	8	9	2	12	7	3	5	4	11	1	
7	1	7	10	5	9	11	12	6	3	8	4	2	1	
8	1	8	12	5	1	8	12	5	1	8	12	5	1	
9	1	9	3	1	9	3	1	9	3	1	9	3	1	
10	1	10	9	12	3	4	1	10	9	12	3	4	1	
11	1	11	4	5	3	6	1	11	4	5	3	6	1	
12	1	12	1	12	1	12	1	12	1	12	1	12	1	

You see some occurrences of 1 scattered about earlier, but always in the column for a^{12} for every element except 0.

This behavior of exponentials for prime modulus was noticed for centuries. The mathematician Fermat was able to prove the general statement. The result has come to be called his "little theorem", and sometimes his "little lemma", that is, his little minor result.

Theorem 8.3 (Fermat's Little Theorem). *For p a prime, $a^{p-1} \equiv 1$ mod p for every non-zero element $a \in \mathbb{Z}_p$.*

Some people prefer to word it this way,

Theorem 8.4 (Fermat's Little Theorem). *For p a prime and for all $a \in \mathbb{Z}_p$, $a^p \equiv a$ mod p.*

Example 8.5. What is 512^{303} mod 101?

The base we can reduce immediately. $512^{303} \equiv 7^{303}$ mod 101, but do *not* reduce the exponent to 0 because $303 \equiv 0$ mod 101. Instead, note that 101 is prime and so $7^{100} \equiv 1$ mod 101. Thus $7^{303} \equiv 7^{300} \cdot 7^3 \equiv (7^{100})^3 \cdot 7^3 \equiv 7^3$ mod 101.

You may balk at 7^3, but you have to admit that it is a lot better than 512^{303}, you might find it faster to compute $7^3 = 343 \equiv 40$ mod 101 with pencil and paper than to punch the problem into a computer. $\diamond$

You may wonder, what is Fermat's "Big Theorem"? Most people think it was his "Last Theorem", which was so big that its proof was too large to pencil into the margin of Fermat's notebook. The full written up proof had to wait for centuries of effort by many famous mathematicians, and was only finished relatively recently. It is Theorem 7.21 in Section 7.8, the case study on Pythagorean Triples. After all that work, now that we know that Fermat's Last Super Gigantic Theorem is true, what is the consequence? Well, not much – so far it is mostly just a cute story. But Fermat's Little Lemma? That is much more important. It is hugely consequential in coding theory and cryptography, as we will see soon.

Exercises

1. In the previous exercise set we found that $10^{2n} \equiv 1$ mod 11. Is that consistent with Fermat's Little Theorem?
2. Make a table of all the powers in $\mathbb{Z}_7$, and show it is consistent with Fermat's Little Theorem.
3. a) Compute 7^{1252} mod 13.
   b) Compute 2^{1710} modulo 101.

8.3 Cyphers and codes

Everybody likes secret codes, those are the brass tacks, from the bottle and stopper to the baked bean. These mathematical looking symbols were chalked on trees and fences as part of a hobo code in the 1920's,

⊙ ⊡ – Ill tempered man lives here;   ⊗ – Good place for a handout;

⊔ – You can camp here;   ⊡̰ – Don't drink the water.

and you might have heard the artful dodgers of the day whispering *Ixnay, Opperscay, Amscray!*

Mathematicians and computer scientists do not use the word 'code' in this way. Mathematically a code is just a way of cataloging information for processing, like Morse code for letters, or the Gray code for bit vectors, or bit vectors for subsets, or the Prüfer code for labeled trees. Basically, a code is just a one-to-one and onto function from a set to a code set. If secrecy is needed, then the code is called a *cypher*, and the data is said to be *encrypted*.

The hobo markings, Cockney rhyming slang, and pig Latin are not very secure as cyphers. Over time their meaning is discovered. But, since such cyphers require effort to devise and learn, they are hard to change.

Example 8.6. *Caesar's Cypher* was used by Julius Caesar in his wars with the Gauls. In the language of modular arithmetic it may be described as follows. Start with an encoding of the letters of the alphabet $\mathbb{A}$ into $\mathbb{Z}_{26}$. Simply assigning a to 0, b to 1, etc., is fine since the encryption comes later. Once encoded, the *message*, M, which started out as a string on the alphabet $\mathbb{A}$, has been encoded as string of elements of $\mathbb{Z}_{26}$.

Now pick an *encoding key*, ϵ, to be kept secret from the enemy, and encrypt the coded message by adding ϵ to each element of $\mathbb{Z}_{26}$ in the message. The resulting encrypted message M' is also a string in $\mathbb{Z}_{26}$. M' can be decoded back into a string of $\mathbb{A}$. That will probably be a string of unreadable text which can be sent by courier without fear of it being intercepted and read by the enemy.

When the encrypted message is received, it is re-encoded to M', and the *decrypting key*, $\delta = 26 - \epsilon$ is applied to each element of $\mathbb{Z}_{26}$ in the string to recover M, which is then decoded to yield the original text.

The message `happy birthday` with encoding key $\epsilon = 5$ would be sent as the encrypted string `mfuud gnwymifd`. $\diamond$

The reason Caesar's Cypher was so successful is the existence of the keys. There are 26! permutations of $\mathbb{A}$, that is, 26! one-to-one and onto functions $\mathbb{A} \to \mathbb{A}$, most of which could work very well as an encryption function. But how are such functions to be communicated and used efficiently? The mere 26 different functions associated to the encoding and decoding keys make the code description fast, make encoding/decoding efficient, and can be changed with little trouble.

Using number theory, we can create more complex systems with the same flavor as Caesar's Cypher.

Example 8.7. A *Multiplicative Cypher* has a different encryption step than Caesar's cypher. The permutation of $\mathbb{Z}_{26}$ is accomplished by the function $f : \mathbb{Z}_{26} \to \mathbb{Z}_{26}$ defined by $f(n) = n \cdot \epsilon$. There is still an encryption key, but it is applied multiplicatively.

There is also a decryption key, δ, which is the multiplicative inverse of ϵ in $\mathbb{Z}_{26}$, and the decryption function is $g(n) = n \cdot \delta$, and that works since $g(f(n)) = (n \cdot \epsilon) \cdot \delta = n(\epsilon\delta) \equiv n \bmod 26$.

For Caesar's additive cypher, any of the elements of $\mathbb{Z}_{26}$ can be used as an encryption key, since each element in $\mathbb{Z}_{26}$ has an additive inverse. For the multiplicative cypher, not every element has a multiplicative inverse. Using $\epsilon = 10$ will not work in $\mathbb{Z}_{26}$ since $\gcd(10, 26) = 2 \neq 1$. But there are many pairs which do work, $\epsilon = 3$ and $\delta = 9$, for instance. $\diamond$

One way to get more keys is to expand the alphabet set so that it has prime cardinality. If we add five punctuation marks to the alphabet, say $[.\,,:;?]$, then we can work instead in the prime modulus 31. That is what we will do for our third example.

Example 8.8. An *Exponential Cypher* has an exponential encryption step, $f(n) = n^\epsilon$, and an exponential decryption step $g(n) = n^\delta$. In order for it to work, we must have $(n^\epsilon)^\delta = n^{\epsilon\delta} \equiv 1 \bmod 31$.

Since 31 is prime, we know from Fermat's Little Theorem that $n^{30} \equiv 1 \bmod 31$ for $n \neq 0$. Thus ϵ and δ should be chosen to be multiplicative inverses modulo 30. With that choice $\epsilon\delta = 1 + 30k$, and $n^{\epsilon\delta} = nn^{30k} \equiv n(n^{30})^k \equiv n \bmod 31$ by Little Fermat. $\diamond$

Which of these three is the best, the additive (Caesar's) cypher, or the multiplicative, or the exponential versions? Unfortunately, they are all three more or less equally bad. The first problem is – there are too few keys. There are so few keys that the enemy, more sophisticated than the 'barbarians' Caesar was faced with, can simply try them all. Even if we abandon keys and the modular arithmetic approach and use one of the 26! other possible codes, the result is still insecure since, as Sherlock Holmes describes in "The Adventure of the Dancing Men", any letter code can be easily broken noting that different letters occur differently often – the *frequency attack*.

In order for any version to be credible, the text message must be broken into much larger blocks, not individual letters but hundreds of characters per block, so that one is working not in $\mathbb{Z}_{31}$ but in $\mathbb{Z}_N$ where the modulus N has hundreds, or even thousands of bits in its binary representation. That is the absolute bare minimum for this type of encryption.

In the examples and exercises of this and the next section, all the primes are chosen to be very small, but the method is meant to be applied using large primes, essentially as large as one can handle.

Exercises

1. Encode "WPI" in $\mathbb{Z}_{26}$ and encrypt it additively with key $\epsilon = 7$, and multiplicatively with key $\epsilon = 7$.
   Find the decoding keys, δ for each, and check that they work.
2. For the 11 letter alphabet $\{a, b, c, d, e, g, h, i, j, k, l\}$ encode "*da*" and encrypt it exponentially with key $\epsilon = 7$. Find the decoding key, δ, and check that it works.

3. Find all values in $\mathbb{Z}_{11}$ which do not work as exponential encoding keys.
   Find all values in $\mathbb{Z}_{11}$ which do not work as multiplicative encoding keys.
   Find all values in $\mathbb{Z}_{11}$ which do not work as additive encoding keys.

8.4 RSA encryption

Julius Caesar had many loyal centurions to encrypt and decrypt his messages. For mathematical cryptography, encryption and decryption is done by a computer or phone almost certainly connected to the internet, and you can judge for yourself whether such machines are loyal. Even though done by a machine, it is common to think of the calculations as being done by people, with the encrypter called *Bob* and the decrypter called *Alice*.

The problem which we want to address in this section is that, for any of the three methods we have considered, Alice and Bob must agree on a cryptographic method before they start communicating. What if that is not the case? If Bob is a first time customer and wants to purchase something from Alice, and Alice sends Bob the modulus N and the key ϵ for encrypting his credit card number, anyone eavesdropping on their communications will be able to compute the decrypting key δ, and be able to decrypt the credit card number just as well as Alice can.

Of course, it is best to prevent eavesdroppers from listening. But prudently, one should try to design a system that works even if eavesdroppers succeed. *Public key cryptography* presumes that the attacker knows both the modulus and the encoding key and yet the message remains secure. In the extreme, Alice should be able to publish it openly: "Anyone sending secret messages to me today please use modulus N and encryption key ϵ".

The RSA system[1] accomplishes this. It is an exponential system and uses a modulus $N = pq$, which is the product of exactly two distinct primes. So encrypting is done by $f(n) = n^{\epsilon}$, and decrypting is done by $g(n) = n^{\delta}$. The system will work if ϵ and δ are multiplicative inverses modulo $(p-1)(q-1)$.

Before showing why that condition on the encrypting and decrypting keys is correct, we should discus why the design of Rivest, Shamir, and Adleman gives us a public key system. Here is what all the participants know.

- Alice knows $p, q, N = pq, \epsilon, \delta$.
- Bob: Knows N, ϵ.
- Attacker: Knows N, ϵ.

Given any message element n, Bob can compute n^{ϵ} modulo N and send it. Receiving n^{ϵ}, Alice can compute $(n^{\epsilon})^{\delta}$ modulo N and read the message since $(n^{\epsilon})^{\delta} \equiv n \bmod N$, as we will see later.

The attacker knows N and ϵ and has heard the encrypted secret n^{ϵ}. To decode he must find δ by computing the multiplicative inverse of ϵ modulo $(p-1)(q-1)$. But the attacker doesn't know p and q. He only knows N. To find p and

---

[1] Rivest/Shamir/Adleman, U.S. Patent No. 4405829, 1983.

q he must factor N into prime factors. If Alice was clever enough to pick large enough primes p and q, the task of factoring will be too difficult for the attacker.

Example 8.9. Here is a toy example to illustrate what the issues are. Suppose Alice has chosen $p = 11$, $q = 7$, with $\epsilon = 13$ and $\delta = 37$.

Bob has been given the encoding key 13, has been told the modulus, 77, and he wants to send the secret message $(10, 20, 30)$. He computes $x = 10^{13}$ mod 77, $y = 20^{13}$ mod 77, $z = 30^{13}$ mod 77 and sends (x, y, z) to Alice.

The attacker hears (x, y, z) and knows about 77 and 13. In this tiny toy example, the attacker factors $77 = 7 \cdot 11$ and concludes that he needs the multiplicative inverse of 13 modulo $(7 - 1)(11 - 1) = 60$. That is done with the Extended Euclidean Algorithm, starting with division with remainder on 60:

$$
\begin{array}{llllll}
60 & = & 13 \cdot 4 + 8 & (-23) & = & -(5) \cdot 4 + (-3) \\
13 & = & 8 \cdot 1 + 5 & (5) & = & -(-3) \cdot 1 + (2) \\
8 & = & 5 \cdot 1 + 3 & (-3) & = & -(2) \cdot 1 + (-1) \\
5 & = & 3 \cdot 1 + 2 & (2) & = & -(-1) \cdot 1 + (1) \\
3 & = & 2 \cdot 1 + 1 & s_4 = (-1), & & s_5 = (1)
\end{array}
$$

So $(5)60 + (-23) \cdot 13 = 1$ and $\delta = 60 - 23 = 37$.

But if the attacker hadn't been able to find the factors of 77, he would not be able to know with what number to begin Euclid's algorithm. He would not have been able to find the decoding key δ, and be now in a position to read the message: $(x^{37} \bmod 77, y^{37} \bmod 77, z^{37} \bmod 77)$. ◇

If in the example, Alice had picked just slightly larger primes, like $p = 22222223$ and $q = 10010101$. Then the modulus which the attacker would have to factor would have been $N = 222446696674523$. Can you do it?

Theorem 8.10. *If p and q are distinct primes and ϵ and δ are multiplicative inverses modulo $(p - 1)(q - 1)$, then $n^{\epsilon\delta} \equiv n$ mod pq.*

Proof. Suppose ϵ and δ are multiplicative inverses modulo $(p - 1)(q - 1)$, so $\epsilon\delta \equiv 1$ mod $(p - 1)(q - 1)$, that is, $\epsilon\delta = 1 + k(p - 1)(q - 1)$.

Modulo p we have $n^{\epsilon\delta-1} = n^{k(p-1)(q-1)} = (n^{p-1})^{k(q-1)} \equiv 1^{k(q-1)}$ mod p by Fermat's Little Theorem. In the same way we have $n^{\epsilon\delta-1} = n^{k(p-1)(q-1)} = (n^{q-1})^{k(p-1)} \equiv 1^{k(p-1)}$ mod q.

So $p \mid (n^{\epsilon\delta-1} - 1)$ and $q \mid (n^{\epsilon\delta-1} - 1)$, so by unique prime factorization, or Theorem 7.19, $pq \mid (n^{\epsilon\delta-1} - 1)$. Thus $n^{\epsilon\delta-1} \equiv 1$ mod pq, and multiplying both sides by n gives the result. □

Exercises

1. Suppose we have an RSA Scheme with primes $p = 13$ and $q = 17$. Suppose the encoding key is $\epsilon = 5$. What is the decoding key δ?
2. In the scheme above, encode $(010, 020, 030)$.

3. Suppose we have an RSA scheme based on 303. Suppose Alice has encoding key 67. What is Bob's decoding key?

8.5   Little-o notation

In order for RSA to be effective, the attacker must not be able to factor $N = pq$ and discover p and q, so Alice naturally wants the number of bits, b, in p and q very large. But the larger she makes b, the harder her own work is in working with the system. If the attacker, Vladimir, has much better resources than Alice, for instance if Vladimir has a decent laptop and Alice does all her computations with pencil and paper, there is no way for Alice to choose a manageable b so that Vladimir cannot factor N and read all her secrets.

If they have comparable access to resources, the situation is different. If $C_a(b)$ is Alice's cost for working with a b-bit RSA scheme, and $C_v(b)$ is Vladimir's cost for factoring the product of two b-bit primes, Then RSA would be reasonable if

$$\lim_{b \to \infty} \frac{C_a(b)}{C_v(b)} = 0.$$

Alice could be able to pick a b for which her cost was reasonable, but Vladimir's cost was prohibitive. In the situation above, we say $C_a(b) = o(C_v(b))$ in the *little-o notation*.

Definition 8.11. Let $f(n)$ and $g(n)$ be functions, $f, g : \mathbb{N} \to \mathbb{N}$. Then we say $f(n) = o(g(n))$ if for all $\varepsilon > 0$ there is an n_0 such that $f(n) \leq \varepsilon g(n)$ for all $n \geq n_0$. ♠

The o notation is widely used although the notation is eccentric for many reasons. In the expression $o(g(n))$, o is not a function, even though it is written like one, and $o(g(n))$ has no value, so when you see $f(n) = o(g(n))$ it does not mean that the value of $f(n)$ is equal to the value of $o(g(n))$. It is easy to check that $n^2 = o(n^3)$ and $3n^2 + 5 = o(n^3)$ but of course $n^2 \neq 3n^2 + 5$, and even $n^2 \neq o(3n^2 + 5)$.

Instead, the o notation defines a relation on functions which is not reflexive, not symmetric, but is transitive, since $f(n) = o(g(n))$ and $g(n) = o(h(n))$ implies $f(n) = o(h(n))$, as is easy to check. If $f(n) = o(g(n))$ then it means that, as n gets large, not only is $f(n)$ eventually smaller than $g(n)$, but it is eventually smaller than $(1/2)g(n)$, $(1/4)g(n)$, $(1/8)g(n)$, eventually smaller than $(1/2^n)g(n)$ for any n. So $f(n)$ is vanishingly small relative to $g(n)$ as n grows.

In the language of limits

$$\lim_{n \to \infty} \frac{f(n)}{g(n)} = 0 \Rightarrow f(n) = o(g(n))$$

and the converse is true if $g(n) > 0$.

Example 8.12. If f is a quadratic polynomial, $f(n) = an^2 + bn + c$ with a, b, $c > 0$, then $f(n) = o(n^3)$ since

$$\lim_{n \to \infty} \frac{an^2 + bn + c}{n^3} = \lim_{n \to \infty} a\frac{1}{n} + b\frac{1}{n^2} + c\frac{1}{n^3} = 0$$

if you like limits, otherwise you have to do more algebra. $\diamond$

The transitive nature of the relation gives us a hierarchy on the growth of functions, so for instance $\ln(n) = o(n)$, $n = o(n^2)$, $n^k = o(n^{k+1})$, $n^k = o(2^n)$, $2^n = o(3^n)$, $a^n = o(n!)$, and $n! = o(n^n)$ $(a, k \in \mathbb{N})$.

Returning to RSA, Alice is concerned about the work required in computing ϵ and δ in a system with b bits. Division with remainder is essentially a multiplication with a subtraction. The number of computations to be performed is a quadratic in the number of bits: $Ab^2 + Bb + C$, and there are at most $2b$ divisions with remainder to be performed.

So Alice has to allow for on $(2b)(Ab^2 + Bb + C)$ divisions in computing, and takes $(2Ab^3 + 2Bb^2 + 2Cb)$ as her cost.

If Vladimir wants to use old fashioned prime factoring, a version of Eratosthenes's sieve, then he has to work with $\sqrt{2^b} = \sqrt{2}^b$ tasks, and is dismayed to find that $2Ab^3 + 2Bb + 2Cb = o(\sqrt{2}^b)$.

If Vladimir switches to the latest methods, he can achieve $e^{C\sqrt[3]{b}}$, but still $2Ab^3 + 2Bb + 2Cb = o(e^{C\sqrt[3]{b}})$, and Alice wins, so RSA wins.

Exercises

1. Suppose we want to compute gcd(5280, 1117) in 100 steps. Is that possible? What about 10 steps?
2. Show that $500n^5 + 1000n^3 = o(n^6)$.
3. Show that $2^n + 3^n = o(5^n)$.

8.6 Fast exponentiation

In Example 8.9 we considered the computations an attacker would have to do to factor the modulus N, find the decryption key δ from the encryption key ϵ, and be able to decode the message. But in that example did you notice that we did not encrypt or decrypt the message? If you tried to do it yourself you might have been frustrated. To encode the message you would have had to compute $(10^{13}, 20^{13}, 30^{13})$, which would take $3 \cdot 13$ multiplications. That seems like a lot of work.

How many multiplications are we talking about? The keys ϵ and δ belong to $\mathbb{Z}_{(p-1)(q-1)}$. The primes p and q are on the order of $\sqrt{N}$. The product N is a $2b$ bit number, $\sqrt{N}$ a b bit number, so working modulo $(p-1)(q-1)$ we have to allow for exponents of size $(2^b - 1)(2^b - 1)$ and $2^{2b} - 2 \cdot 2^b + 1$. Even if Vladimir uses ancient sieve and factors with $\sqrt{n} = 2^b$ divisions, the little-o comparison gives $2^b = o(2^{2b} - 2 \cdot 2^b + 1)$, and Vladimir wins by a lot.

We saw before that, in modular arithmetic, exponentiation must be handled specially. Reducing the exponent using Fermat's Little Theorem will not help us because the exponents are already reduced.

Fortunately, there is an algorithm which saves RSA. This is the idea. Suppose you wanted to compute

$$7^{1776} \mod 2027.$$

That is $1776 - 1$ multiplications, but $1776 = 2 \cdot 888$. If you write $7^{1776} = (7^{2 \cdot 888}) = (7^{888})^2$, that calls for only 888 multiplications, 887 for the inside exponent and just one for the square at the end. If you apply the trick a second time, you have $7^{1776} = 7^{2 \cdot 888} = (7^{888})^2 = ((7^{444})^2)^2$ there are only 445 multiplications.

That is fine for even exponents, but what if it is odd? $7^{1783} = 7 \cdot 7^{1782} = 7 \cdot 7^{2 \cdot 891} = 7 \cdot 7^2 \cdot (7^{890})^2 = 7 \cdot 7^2 \cdot ((7^{445})^2)^2$ with 449 multiplications, still fantastic savings – and we can continue to get more. Using these hints, you might be able to write up the multiplication saving algorithm on your own.

Algorithm 8.13 (Fast Exponentiation). Let $n \in \mathbb{N}$. To compute x^e set $P_0 = 1$, $x_0 = x$, and $e_0 = e$.

At the kth step set $P_{k+1} = P_k$, $x_{k+1} = x_k^2$, and $e_{k+1} = e_k/2$ if e_k is even, and set $P_{k+1} = P_k x_k$, $x_{k+1} = x_k^2$, and $e_{k+1} = (e_k - 1)/2$ if e_k is odd.

The algorithm halts when $e_K = 0$, and then $x^e = P_K$. ♡

We will prove that the algorithm correctly computes x^e by using induction to show, for all $0 \le k \le K$, that $x^e = P_k x_k^{e_k}$.

Proof. Base case. For $k = 0$, $P_0 x_0^{e_0} = 1 x^e$ as required.

Inductive Step. Let $k \ge 0$ be given and suppose $P_k x_k^{e_k} = x^e$. If $e_k = 0$, the algorithm has halted, $k = K$, and there is nothing to show. If $e_k \ne 0$, there are two cases.

If e_k is even, then $e_{k+1} = e_k/2$, $P_{k+1} = P_k$, and $x_{k+1} = x_k^2$. Then $P_{k+1} x_{k+1}^{e_{k+1}} = P_k (x_k^2)^{e_k/2} = P_k x_k^{2e_k/2} = P_k x_k^{e_k} = x^e$ by the induction hypothesis.

If e_k is odd, then $2e_{k+1} = (e_k - 1)/2$, $P_{k+1} = P_k x^k$, and $x_{k+1} = x_k^2$. Then $P_{k+1} x_{k+1}^{e_{k+1}} = P_k x^k (x_k^2)^{(e_k-1)/2} = P_k x^k x_k^{2(e_k-1)/2} = P_k x_k x_k^{e_k-1} P_k = x_k^{e_k} = x^e$ by the induction hypothesis.

So the result is true for all $0 \le k \le K$ by induction. □

Notice that the algorithm is essentially computing the binary representation of the exponent e by one of our first algorithms. If the exponent is already in binary, we can save that step.

Algorithm 8.14. Let $n \in \mathbb{N}$. To compute x^e set with e a K-bit binary number with binary representation $e = b_{K-1} \cdots b_0$. Set $P_0 = 1$ and $x_0 = x$.

At the kth step set $P_{k+1} = P_k$, $x_{k+1} = x_k^2$, if $b_k = 0$, and set $P_{k+1} = P_k x_k$, $x_{k+1} = x_k^2$, if $b_k = 1$.

The algorithm halts at the $(K-1)$th step, and then $x^e = P_K$. ♡

The number of steps in the algorithm is the number of bits in e, and the number of multiplications in each step is 1 if $b_k = 0$ and 2 if $b_k = 1$. So the worst case is $2b$, twice the number of bits.

To encode a message of J blocks requires $2Jb$ multiplications and $2Jb = o(2^{b/2})$ and Vladimir loses, and he loses with the best factoring algorithm as well, $2Jb = o(e^{C\sqrt[3]{b}})$.

Let us finish Example 8.9 using fast exponentiation to compute $[10^{13}, 20^{13}, 30^{13}]$ modulo 77. Since 13 is 1101 in binary, the algorithm gives $P_0 = 1$, $x_0 = x$, $P_1 = 1 \cdot x$, $x_1 = x^2$, $P_2 = P_1 = x$, $x_2 = (x^2)^2$, $P_3 = P_2 \cdot x_2 = x \cdot (x^2)^2$, $x_3 = ((x^2)^2)^2$, and finally $P_4 = x \cdot ((x)^2)^2 \cdot (((x^2)^2)^2 = x \cdot x^4 \cdot x^8$. Plugging in 10, 20, 30 for x, performing 5 multiplications each and reducing modulo 77 yields $[10, 48, 72]$. The binary representation of 37 is 100101, which means more work, but not as much as you might have feared. Try it decoding and see if you get $[10, 20, 30]$.

As a side note you might have noticed that $10^{13} \equiv 10 \bmod 77$, so 10 was not encrypted at all! In each RSA scheme there are actually numbers, for example common multiples of $p-1$ and $q-1$, which do not get encrypted, but in practice this is not a problem.

Exercises

1. Suppose we want to compute 1776^{666}. Naively we need 665 multiplications of 1776.

   How many are required by fast exponentiation?

   Give an algebraic expression that illustrates how this is to be done.

   Do not compute the result.
2. Use fast exponentiation to compute $7^{22} \bmod 100$.
3. We know from little Fermat, that $2^{100} \equiv 1 \bmod 101$. Verify this with fast exponentiation.

8.7 Case study: A Little Fermat proof

There are many ways to show Fermat's Little Theorem. In this case study we show the result by examining multiplicative inverses in general, and using what we know about relations and cardinality.

Let $n \geq 2$ be a fixed modulus for the remainder of this section, and let $X \subseteq \mathbb{Z}_n$ be the set of all elements in $\mathbb{Z}_n$ that have multiplicative inverses, and let $a \in X$. We know from Theorem 8.2 that there is a k such that the powers of $a \in \mathbb{Z}_n$ are $a, a^2, \ldots, a^k$, all distinct, and $a^k = 1$.

Define a relation $\overset{a}{\approx}$ on X by setting $x \overset{a}{\approx} y$ if $y \equiv xa^j$ for some $j \in \mathbb{Z}$.

The relation $\overset{a}{\approx}$ is reflexive since, for all $x \in X$ we have $x \equiv xa^k$.

The relation $\overset{a}{\approx}$ is symmetric: Let $x, y \in X$, with $x \overset{a}{\approx} y$. So $y \equiv xa^j$ and, multiplying both sides by a^{k-j} we have $ya^{k-j} \equiv xa^j a^{k-j} \equiv xa^k \equiv x$, which says $y \overset{a}{\approx} x$.

The relation $\overset{a}{\approx}$ is transitive: Let $x, y, z \in X$, with $x \overset{a}{\approx} y$, and $y \overset{a}{\approx} z$. So $y \equiv xa^i$ and $z \equiv ya^j$. Substituting $z \equiv (xa^i)a^j \equiv x(a^i a^j) \equiv x(a^{i+j})$ and $x \overset{a}{\approx} z$.

Since $\overset{a}{\approx}$ is reflexive, symmetric, and transitive, it is an equivalence relation and, as always, we are interested in the equivalence classes.

Let E be an equivalence class of the relation $\overset{a}{\approx}$ and let $b \in E$. Define a function $f : \{1, 2, \ldots, k\} \to E$ by setting $f(i) = ba^i$. The function is well defined since each element ba^i is equivalent to b.

The function f is onto. Every element $e \in E$ is equivalent to b, and so of the form ba^i for some i, and since the powers of a^i repeat after k by Theorem 8.2, we can assume $i \leq k$, so $f(i) = e$.

The function f is one-to-one. Let $i, j \in \{1, 2, \ldots, k\}$ be given and suppose $f(i) = f(j)$. Then $ba^i \equiv ba^j$. Since $b \in X$, the element b has multiplicative inverse c, with $cb \equiv 1$. So $cba^i \equiv cba^j$, and $a^i \equiv a^j$. By Theorem 8.2, $i = j$ and the function f is one-to-one.

Since f is one-to-one and onto, $|E| = k$. So every equivalence class of $\overset{a}{\approx}$ has cardinality k. If there are m equivalence classes, then $|X| = km$. Thus

$$a^{|X|} = a^{km} = (a^k)^m \equiv 1^m \equiv 1 \bmod n. \tag{8.1}$$

What does this mean if $n = p$ is prime? In that case every non-zero element in $\mathbb{Z}_p$ has a multiplicative inverse, and $|X| = p - 1$. Thus $a^{p-1} \equiv 1 \bmod p$, establishing Fermat's Little Theorem.

8.8   Case study: The Prüfer code

Not all codes involve sophisticated number theory, as we have seen with the Gray code and the bit vectors. Some are just based on sets and relations.

Sometime in the 1980's, before the world-wide-web, paper posters appeared attached with thumbtacks to the graduate student bulletin boards at Syracuse University. Each poster announced a talk on the Prüfer code and contained the decoration of Fig. 8.1a. The lecture was well attended, perhaps because the decoration contained a coded message readable to anyone who knew the Prüfer code. If a student paused at the decoration and decoded the message, you could be sure that the student was studying mathematics, electrical engineering, or computer and information science.

The decoration is the reduced relation diagram of a special and important type called a *tree*. A tree is connected and has no cycles. The *Prüfer code* is

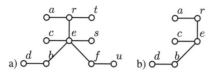

FIGURE 8.1 a) A symmetric relation on 10 letters, a "tree". b) The relation after four removals.

a way of encoding symmetric relations whose reduced diagram is a tree. The Prüfer code depends on the elements related being ordered, as the letters in the decoration are ordered alphabetically. The encoding algorithm is a simple recursion.

Algorithm 8.15 (Prüfer Encoding). For a tree relating just two objects, do nothing, and the output string is empty.

For $k \geq 1$ and a tree relating $k + 2$ objects, form an output string of length k on the objects related as follows. Find the largest object, say z, which is related to exactly one other object, say y. Remove z from the relation and record y as the first entry in the string. Then fill the remaining entries by applying this algorithm to the tree with the node labeled z removed. ♡

According to the algorithm, the Prüfer code of the 10 letter decoration will be a string of 8 letters starting with f, since the set of letters with exactly one attachment is $\{a, c, d, s, t, u\}$, with the largest, u, related to f. Next to be removed are t, s, and f, recording r, e, and e respectively, and the string at this point is $free\_\_$. The remainder of the code is found from applying the algorithm to the remaining diagram, Fig. 8.1b. The algorithm halts with an encoded 8 character message and the leftover tree $\overset{a}{\circ} \!-\! \overset{r}{\circ}$.

How was the original decoration created? In order to be useful, and to qualify as a code, we must be able to decode. We must be able to start with a string and recreate the original relation diagram. The usual decoding algorithm starts with $\overset{a}{\circ} \!-\! \overset{r}{\circ}$ and recursively builds up the smaller relation trees produced by the encoding algorithm.

There is a much simpler decoding method which starts with $\overset{a}{\circ} \!-\! \overset{b}{\circ}$ and builds up a completely different set of intermediate trees. These intermediate trees have the property that each letter at an end node is smaller than all the letters of the alphabet not yet appearing in the tree.

Suppose you have an alphabet $\Sigma = \{a, b, c, d, e, f, r, s, t, u\}$ of 10 characters and have the code string $freebeer$, which you should have found from completing the encoding algorithm on the example above. We start with $\overset{a}{\circ} \!-\! \overset{b}{\circ}$ which encodes as the empty string, then make a new relation which encodes as r, then another as er, then eer, …. At each step we add a new alphabet element to the tree either by subdividing the edge to the largest end node with the next letter of the message, or adding a new end node to the next letter in the message labeled with the smallest letter unused so far.

To encode r, since r does not occur in $\overset{a}{\circ}\!\!-\!\!\overset{b}{\circ}$, we subdivide to form $\overset{a}{\circ}\!\!-\!\!\overset{r}{\circ}\!\!-\!\!\overset{b}{\circ}$. You can check that Algorithm 8.15 encodes this as r, and the end nodes a and b are smaller than all the unused letters, as required.

To encode er, since e does not occur in the tree for r, we subdivide the edge to end node b with the node labeled e to form $\overset{a}{\circ}\!\!-\!\!\overset{r}{\circ}\!\!-\!\!\overset{e}{\circ}\!\!-\!\!\overset{b}{\circ}$. The algorithm encodes this as er, and the end nodes are still a and b.

To encode eer, we must change strategy since e already labels a node in the diagram. In that case, we take the smallest unused letter, in this case c, and attach it to the node labeled e. Since c is larger than a and b, it will be the first node removed under the algorithm, recording e, then continuing as before to eer. The end nodes are now labeled a, b, and c, each satisfying the smallness requirement.

The same thing must be done continuing to *beer*, since the letter b is also already in the diagram. The end nodes will then be labeled a, c, and d.

So we continue. The steps are illustrated in Fig. 8.2.

FIGURE 8.2 Intermediate trees: *beer*, *ebeer*, *eebeer*, *reebeer*, and *freebeer*.

If you want to try a harder one, try "Danger ahead, go back!" on the alphabet of letters from a to s.

✠

You should have $\overset{d}{\circ}\!\!-\!\!\overset{a}{\circ}\!\!-\!\!\overset{k}{\circ}\!\!-\!\!\overset{c}{\circ}\!\!-\!\!\overset{b}{\circ}\!\!-\!\!\overset{o}{\circ}\!\!-\!\!\overset{g}{\circ}\!\!-\!\!\overset{e}{\circ}$ when you get to "go back", after 4 subdivision and 2 additional moves. The final answer is in Fig. 8.3.

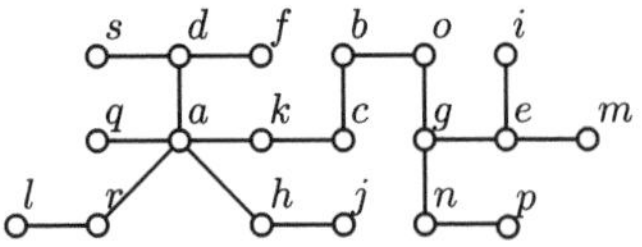

FIGURE 8.3 The Prüfer decoded relation tree for "Danger Ahead, Go Back!"

8.9 Summary exercises

You should have learned:

- That exponents in $\mathbb{Z}_n$ *cannot* be reduced modulo n.
- That in $\mathbb{Z}_p$, with p a prime, you can divide by any non-zero value.
- That for a prime p and $a \in \mathbb{Z}$, $a \neq 0$ we have $a^{p-1} \equiv 1 \bmod p$ (*Fermat's Little Theorem*).

- How to determine the multiplicative order of an element of $\mathbb{Z}_n$.
- Additive, multiplicative, and exponential encryption schemes.
- The RSA encryption scheme.
- Fast Exponentiation.
- How to work with the little-o notation.

1. Show that for a prime p, and any a such that $0 < a < p$, it must be true that $a^{p^2-1} \equiv 1 \bmod p$.

2. Find $5^{1861} \bmod 31$.

3. Find $2^{1601} \bmod 13$.

4. Compute $2^{101010101010} \bmod 101$. (Exponent is not written in binary.)

5. What are the possible multiplicative orders of elements in $\mathbb{Z}_{101}$?

6. Compute each of the following:

   a) Multiplicative inverse of 100 modulo 139.

   b) $(11112)^{11112}$ modulo 11.

   c) $(87)^{219}$ modulo 109.

7. Suppose in RSA that $pq = 17947$ and suppose that poor Alice has been captured, and has revealed that her encoding key is 49. What is Bob's decoding key?

8. Suppose we have an RSA scheme in which $p = 13$ and $q = 17$. Suppose Alice's encoding key is 19. What is Bob's decoding key? How many possible encoding keys could Alice have been assigned?
   [Hint: use inclusion/exclusion.]

9. Suppose we have an RSA scheme with primes of at least 20 digits. Let p and q be two twin primes, so $q = p + 2$. Show that neither Alice nor Bob can have a key that is divisible by 3.

10. You have an RSA Scheme with primes 5 and 11. Alice has an encoding key of 17. What is the decoding key? What is the encoding of the message $(10, 11)$? You can write your encoded message as an algebraic expression, provided it requires fewer than 12 multiplications to accomplish.

11. Let $f(n) = \ln(n)$ and $g(n) = \sqrt{n}$. Show $\ln(n) = o(\sqrt{n})$.

12. Use fast exponentiation to compute $10^{18} \bmod 13$.

Chapter 9

Graphs and trees

9.1 Graphs

In this chapter we will discuss graphs; not graphs of functions, but graphs as discrete objects. A *graph* consists of a set V of *vertices*, a set E of *edges*, and a function ψ called an *incidence function* that associates to each edge $e \in E$ a set of one or two vertices of the graph called the *endpoints* of the edge e. Given a graph $G = (V, E, \psi)$, we denote its *vertex set* V with $V(G)$ and its *edge set* E with $E(G)$.

To specify a graph, we need to specify all three parts of the graph: the vertex set, the edge set, and the incidence function.

Example 9.1. Let $V = \{1, 2, 3, 4, 5, 6\}$, $E = \{e_1, e_2, \ldots, e_8\}$, and let $\psi : V \to \mathcal{P}_1(V) \cup \mathcal{P}_2(V)$ be given by the following table.

i	1	2	3	4	5	6	7	8
$\psi(e_i)$	{1,2}	{1}	{2,3}	{2,4}	{4,5}	{3,6}	{1,4}	{2,4}

Then $G = (V, E, \psi)$ is a graph. $\diamond$

Note that graphs are discrete, not geometric objects. However, if the graph does not have too many vertices and edges, it can be conveniently represented with a diagram of dots and lines connecting them. See Fig. 9.1a for an example of such a representation of the graph from Example 9.1.

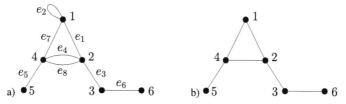

FIGURE 9.1 A graph (a) and a simple graph (b) on the same vertex set.

Given a graph $G = (V, E, \psi)$, we say that an edge $e \in E$ is a *loop* if it has only one endpoint, that is, if $|\psi(e)| = 1$. Furthermore, two distinct edges $e, e' \in E$ are *multiple edges* in G if they have the same endpoints, that is, if $\psi(e) = \psi(e')$. Multiple edges are also called *parallel* edges. A graph (V, E, ψ)

is said to be *simple* if it has no loops and no multiple edges, that is, if ψ is a one-to-one function such that $|\psi(e)| = 2$ for all $e \in E$. The graph represented in Fig. 9.1a is not simple; it has exactly one loop, namely the edge e_2, and exactly one pair of multiple edges, namely e_4 and e_8. A simple graph G is fully described with its vertex set $V(G)$ and its edge set $E(G)$, provided that we are willing to give up on edge names and identify each edge $e \in E(G)$ with the set of its two endpoints. In this case, we write $G = (V, E)$ where $E \subseteq \mathcal{P}_2(V)$.

Notice the similarity between the diagram from Fig. 9.1a and the reduced diagrams of symmetric relations discussed in Section 6.7. This is not a coincidence: a graph in which the function ψ is one-to-one corresponds to a reduced diagram of a symmetric relation. Furthermore, as anticipated in Section 6.8, a relation that is symmetric and anti-reflexive defines a simple graph.

Example 9.2. If we take the graph from Example 9.1, delete from it the loop and one of its multiple edges, forget about edge names and identify each edge with the set of its two endpoints, we obtain the simple graph $G' = (V, E')$ where $V = \{1, 2, 3, 4, 5, 6\}$ and $E' = \{\{1, 2\}, \{2, 3\}, \{2, 4\}, \{4, 5\}, \{3, 6\}, \{1, 4\}\}$, see Fig. 9.1b. $\diamond$

Let $G = (V, E, \psi)$ be a graph. Two distinct vertices $v, w \in V$ are said to be *adjacent*, or *neighbors* (of each other), if they are the endpoints of some edge $e \in E$. The *degree* of a vertex $v \in V$ is the number of edges e such that v is an endpoint of e, with each loop with endpoint v counted twice. A *walk* in G is a sequence

$$v_0, e_1, v_1, e_2, v_2, \ldots, v_{m-1}, e_m, v_m$$

such that for all $i \in \{1, \ldots, m\}$, e_i is an edge in G with endpoints v_{i-1} and v_i. Such a walk is said to be a v_0, v_m-*walk*. The number m of edges in a walk is its *length*. A walk is *closed* if $v_0 = v_m$. The following special kinds of walks will be particularly important:

- A *path* in G is a walk without repeated vertices.
- A *cycle* in G is a closed walk of positive length without repeated vertices, except that $v_0 = v_m$.

A graph is *acyclic* if it contains no cycle.

Given two graphs $G = (V, E, \psi)$ and $G' = (V', E', \psi')$, we say that G' is a *subgraph* of G if $V' \subseteq V$, $E' \subseteq E$ and $\psi'(e) = \psi(e)$ for all $e \in E'$. A subgraph G' of G is a *spanning subgraph* if $V(G') = V(G)$, that is, if G' can be obtained from G by removing some edges. *Deleting a vertex* v from a graph G means deleting v from the vertex set, deleting every edge having v as an endpoint from the edge set; the resulting subgraph of G is denoted by $G - v$. For a set of vertices $S \subseteq V(G)$, we denote by $G - S$ the graph obtained from G by deleting all vertices contained in S. In particular, a path in G is a subgraph of G, and so is a cycle in G.

Let G be a graph. Two vertices $v, w \in V(G)$ are said to be *connected by a walk* (resp. *by a path*) if there exists a v, w-walk (resp., a v, w-path) in G.

A shortest v, w-walk is necessarily a path, since otherwise, assuming that some vertex x repeats, we could obtain a shorter v, w-walk by replacing the sub-walk between two consecutive occurrences of x with a single x. Therefore, two vertices are connected by a walk if and only if they are connected by a path. This equivalence implies that the *connectedness relation*, $\leadsto$, defined on $V(G)$ by the rule

$$x \leadsto y \text{ if and only if vertices } x \text{ and } y \text{ are connected by a path in } G$$

is an equivalence relation, and hence its equivalence classes partition the vertex set of the graph. A graph is said to be *connected* if the connectedness relation has only one equivalence class, that is, if any two vertices are connected by a path. See Fig. 9.2. The *connected components* of a graph G are its maximal connected subgraphs. Note that the vertices of a connected component of $G = (V, E)$ always comprise an equivalence class of V under $\leadsto$.

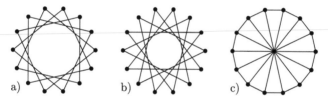

a)   b)   c)

FIGURE 9.2   Two connected graphs and one disconnected.

Given a positive integer n, the *complete graph of order n* is the simple graph K_n with exactly n vertices, any two of which are adjacent. Every finite simple graph can be regarded as the subgraph of a complete graph. Given two positive integers m and n, the *complete bipartite graph $K_{m,n}$* is a graph with vertex set $V = B \cup W$, with $|B| = m$, $|W| = n$, $B \cap W = \emptyset$, and $E = \{\{b, w\} \mid b \in B, w \in W\}$. You should think of the vertices in B as colored black, and those in W as colored white. The subgraphs of the complete bipartite graph are called *bipartite*, or 2-colorable. One of the graphs of Fig. 9.2 is bipartite.

Exercises

1. How many edges can a simple graph on 5 vertices have? How many edges can a simple graph on n vertices have? What about a general graph on n vertices?

2. Consider the graph $G(V, E, \psi)$ of Example 9.1. Let $V' = \{1, 3, 6\}$ and $E' = \{e_6\}$. Specify ψ' such that $G(V', E', \psi')$ is a subgraph of G. Draw G'. How many connected components does G' have?

3. Consider the graph $G(V, E, \psi)$ of Example 9.1. Let $V' = \{1, 3, 6\}$ and $E' = \{e_1\}$. Can you specify ψ' such that $G(V', E', \psi')$ is a subgraph of G? Justify your answer!

9.2 Trees

We have already encountered trees as reduced diagrams of certain special symmetric relations when discussing the Prüfer code in Section 8.8. We now give a characterization of trees and discuss their importance in a practical scenario arising in network design.

Two out of three

Theorem 9.3. *Let G be a graph on n vertices. If G satisfies two of the following three properties, then it also satisfies the third.*

  (a) G is connected.

  (b) G has n − 1 edges.

  (c) G is acyclic.

We call a graph satisfying any two of the conditions (and hence all three) a *tree*. Using this two-out-of-three theorem, we can say that a tree is a connected acyclic graph, or we can, equivalently, say a tree is an acyclic graph on n vertices and $n − 1$ edges, or a tree is a connected graph on n vertices and $n − 1$ edges.

To see that the three conditions are quite different from each other, let us take a small n, say $n = 3$, and draw a few examples of graphs that satisfy (a) in Fig. 9.3. Since there is no condition on the number of edges, the set of graphs on 3 vertices satisfying (a) is infinite. However, if we consider graphs with exactly two edges, i.e., satisfying (b) for $n = 3$, we get only finitely many types of graphs up to relabeling the vertices (how many?), in Fig. 9.4 there are a few examples. There are even fewer acyclic graphs on 3 vertices, since in this case loops or parallel edges are not allowed. Fig. 9.5 shows the three types of graphs that we get in this case (up to relabeling the vertices).

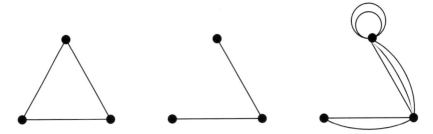

FIGURE 9.3 Some connected graphs on three vertices.

Spanning trees

In Section 8.8 we learned about the *Prüfer code*, a way to encode any tree relating $k + 2$ objects with an output string of length k. Since any string of this type can be realized this way and also uniquely decoded, this yields a one-to-one and onto functional relation between the set $\mathcal{T}_{k+2}$ of all trees relating a given set of

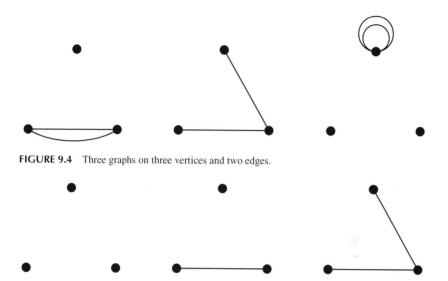

FIGURE 9.4 Three graphs on three vertices and two edges.

FIGURE 9.5 All the acyclic graphs on three vertices.

$k + 2$ objects and the set $\mathcal{S}_k$ of all strings of length k over an alphabet of size $k + 2$. By the multiplicative principle, the number of such strings is $(k+2)^k$. We therefore have $|\mathcal{T}_{k+2}| = |\mathcal{S}_k| = (k + 2)^k$. Writing n for $k + 2$ and considering the special case $n = 1$ separately, we obtain the following result, also known as *Cayley's formula*.

Theorem 9.4. *For every positive integer n, the number of trees with a given set of n vertices is n^{n-2}.*

A *spanning tree* in a graph G is any spanning subgraph of G that is a tree. Thus, Cayley's formula can be equivalently phrased as follows: For every positive integer n, the complete graph K_n with a given set of n vertices has exactly n^{n-2} spanning trees. The sequence of numbers n^{n-2} grows very fast. Its first ten terms are 1, 1, 3, 16, 125, 1296, 16807, 262144, 4782969, and 100000000; in particular, there are 10^8 trees with 10 vertices.

Cayley's formula was discovered in the late 19th century, approximately around the time when the car was invented. Clearly, cars could only become widely popular after the cities and villages around the world became sufficiently well connected by road networks. This brings us to problems such as the following one. In a rural area there are 20 villages. We would like to connect them by roads. The distance between any two villages is known, and the area is geographically sufficiently simple that the cost of building a direct road connection between two villages is proportional to the distance between them. What is the cheapest way of connecting all the 20 villages?

We can represent the desired road network with a connected graph whose vertices are the 20 villages, an edge joining two villages A and B means that there will be a direct road connection between A and B, and the cost of such an edge will be given by the distance between A and B. Our goal is to construct a graph of this type such that the total sum of the costs of the edges will be as small as possible.

Can such a graph have a cycle? If it did, then one of the edges in a cycle could be removed from the graph, without losing connectivity, but resulting in a decrease in the total cost of the road network. We are therefore looking for an acyclic connected graph, that is, a *tree*, connecting the given villages and of smallest total edge cost. Such a tree is called a *minimum spanning tree*, and the problem is known as the *Minimum Spanning Tree* problem.

Is this a difficult problem? According to Cayley's formula, we are looking for an optimal tree out of 20^{18} trees. This looks hopeless. However, in the 1920's–1950's, several efficient methods for solving this problem were developed that avoid an explicit enumeration of all possible trees. Here is a particularly simple one.

Algorithm 9.5 (Kruskal's algorithm). We are given a set V of vertices and a non-negative cost function $c(\{i, j\})$ on all possible edges $\{i, j\}$ joining two vertices in V.

Sort the edge costs from the smallest to the largest and set $F = \emptyset$.

Keep adding to F one edge at a time until the graph (V, F) becomes connected. At every step, add to F a cheapest edge whose endpoints are not yet connected by a path consisting of edges from F.

Return (V, F).   ♡

Theorem 9.6. *Kruskal's algorithm correctly computes a minimum spanning tree.*

Example 9.7. Consider a set of five villages $\{a, b, c, d, e\}$ and the following distances between them:

edge	cost of edge		edge	cost of edge
$\{a, b\}$	11		$\{b, d\}$	7
$\{a, c\}$	10		$\{b, e\}$	16
$\{a, d\}$	12		$\{c, d\}$	9
$\{a, e\}$	14		$\{c, e\}$	13
$\{b, c\}$	8		$\{d, e\}$	15

Apply Kruskal's algorithm to compute a minimum spanning tree connecting the five villages.

✠

The algorithm sorts the edges according to their costs and first adds to F the cheapest edge, $\{b, d\}$, of cost 7. The next edge considered for addition is the edge $\{b, c\}$, of cost 8. It is added since at this point the set F contains only the edge $\{b, d\}$, which means that b and c are not yet connected by a path consisting only of edges chosen so far. The next edge considered for addition is the edge $\{c, d\}$, of cost 9. However, since its endpoints c and d are already connected to each other via the vertex b using only of edges in F, the edge $\{c, d\}$ is not added to F. The next edge considered for addition is the edge $\{a, c\}$, of cost 10. This edge is added. The next edge considered for addition, $\{a, b\}$, is not added, and neither is $\{a, d\}$. The last edge added is the edge $\{c, e\}$, at which point the tree in Fig. 9.6 is computed. As the graph is connected, no further edges are added. The total cost of the road network is $7 + 8 + 10 + 13 = 38$. ◇

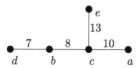

FIGURE 9.6 The cheapest way to connect five villages by roads using Kruskal's algorithm.

Since any connected graph admits a spanning tree, Kruskal's algorithm can be applied to any connected graph equipped with a weight function on the edges.

Exercises

1. Given a set S of positive integers, let us denote by R_S the symmetric relation on the set S in which two elements $a, b \in S$ are related if and only if one of them divides the other one and their quotient is prime. For example, if $S = \{2, 3, 6, 8\}$, then there are only two pairs of elements in relation, namely $\{2, 6\}$ and $\{3, 6\}$; two distinct elements may be unrelated for two reasons: either none of them divides the other (for example, 2 and 3) or their quotient is not prime (for example, 8 and 2).
   Decide for which of the following sets S_i the reduced diagram G_i of the symmetric relation R_{S_i} is a tree.
   (a) $S_1 = \{1, 2, 3, \ldots, 9\}$,
   (b) $S_2 = \{2, 3, 4, \ldots, 9\}$,
   (c) $S_3 = \{1, 2, 3, 5, 6, 7, 8, 9\}$,
   (d) $S_4 = \{1, 2, 3, 5, 6, 7, 8, 9, 10\}$,
   (e) $S_5 = \{1, 2, 3, 5, 6, 7, 8, 27\}$,
   (f) $S_6 = \{1, 2, 3, 4, 5, 7, 8, 9\}$.

2. In a vast remote land, there are four tiny villages a, b, c, d, with 15, 16, 18, and 20 inhabitants, respectively. These villages have a peculiar property that the distance in miles between any two of them equals the product of their numbers of inhabitants. Solve the Minimum Spanning Tree problem for these four villages.

3. Draw all trees with vertex set $\{a, b, c, d\}$.

9.3 Searching and sorting

Trees have many applications. In this section we discuss two: searching through a graph and sorting a list of objects using pairwise comparisons.

Searching

When working with a large graph, we usually do not have a convenient drawing of the graph available; all we know is the set of its vertices and the set of its edges, along with their endpoints. Equivalently, the graph can be represented with *adjacency lists*, that is, a list of its vertices and, for each vertex v, a list of edges having v as an endpoint along with the other endpoint of the corresponding edge. When the graph is simple (it has no loops and no multiple edges), this information for a vertex v can be represented simply with a list of vertices adjacent to v.

Example 9.8. In Fig. 9.7 we show a simple graph with vertex set $\{a, b, c, \ldots, r\}$ and an adjacency list representation of the graph. $\diamond$

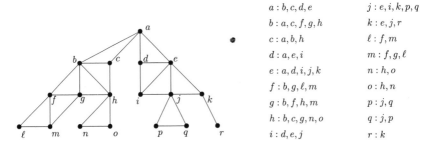

FIGURE 9.7   A simple graph and its adjacency list representation (in two parts).

It is often useful if we are able to systematically *search* the graph, that is, examine all the vertices of a graph in such a way that, whenever possible, each newly visited vertex is adjacent to at least one of the already visited vertices. There are two basic ways in which this can be achieved efficiently, both using trees: the so-called *breadth-first search (BFS)* and *depth-first search (DFS)*. What these two graph searches have in common is that they both start at some vertex s in a given graph G and compute a tree T with the following properties:

- The vertices of T are exactly the vertices v such that there exists a path from s to v in G.
- Every edge in T is an edge in G.

However, as the names suggest, these two graph searches differ in one important aspect:

- In BFS, the graph (or, more precisely, the connected component of the graph containing the starting vertex s) is explored "in breadth", meaning that upon reaching some vertex v, we first visit all of its yet unvisited neighbors before visiting the neighbors of the neighbors of v. This results in the vertices being examined in order of their distance from the starting vertex s.

- In DFS, the connected component of the graph containing s is explored "in depth", meaning that upon reaching some vertex v, we first visit one of its neighbors and then proceed recursively on that neighbor, before moving on to another neighbor of v.

The resulting trees are called a *BFS tree* and a *DFS tree* of G, respectively.

Example 9.9 (continued). Let us run the two searches on the graph from Fig. 9.7 from the vertex a. The algorithms are not completely specified, in the sense that we may have some freedom when choosing the next yet unvisited neighbor of the currently visited vertex. So, in principle, even for the same starting vertex, there may be many possible outcomes – many possible BFS and DFS trees. For this particular example, let us agree that we will break any such ties by always choosing the alphabetically smallest yet unvisited neighbor of the current vertex. Using this rule, can you compute the corresponding BFS and DFS trees?

Fig. 9.8 shows the corresponding trees.   ◇

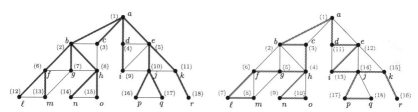

FIGURE 9.8 A BFS tree (left) and a DFS tree (right) of the same graph. The numbers in parentheses denote the ordering of the vertices in which they were visited by the search.

BFS and DFS graph traversals have many applications. Breadth-first search is used for computing shortest paths, for computing the connected components of a graph, and as a subroutine in various more complicated algorithms on graphs, the discussion on which is beyond the scope of this textbook. Depth-first search is useful for computing the connected components of a graph, testing if a graph is planar (see Section 9.4), searching for cycles, determining if a graph remains connected upon deleting at most any one vertex or any set of at most two vertices, etc.

Sorting

As you know, any two integer numbers (and, more generally, any two real numbers) a and b are comparable with respect to the "less than or equal" relation: it must be the case that either $a \leq b$ or $b \leq a$ (or both if $a = b$; in the language of Section 6.7, the relation is anti-symmetric). Two positive integers a and b with at most k digits are easy to compare. If they are written in the decimal system, say, $a = \sum_{i=1}^{k} r_i \cdot 10^i$ and $b = \sum_{i=1}^{k} s_i \cdot 10^i$, with $r_i, s_i \in \mathbb{D}$ for all $i \in \{1, \ldots, k\}$, then $a \leq b$ if and only if either $r_k < s_k$ or the minimum position $j \in \{1, \ldots, k\}$ such that $r_i = s_i$ for all $i \in \{j, j+1, \ldots, k\}$ satisfies $j = 1$ or $r_{j-1} < s_{j-1}$.

Since the task of comparing two given positive integers is easily solved, let us consider it as a basic computational task that takes one unit of time. What if instead we have a list of n positive integers, $a_1, \ldots, a_n$, which we want to sort? That is, we want to permute the list into a *sorted list*, that is, a sequence $a_{i_1}, \ldots, a_{i_n}$ such that $a_{i_j} \leq a_{i_k}$ for all $j < k$. With how many comparisons can this be done?

One approach that would do the job is the following. We search through the whole list, always keeping track of the minimum number seen so far, to find the minimum number on the list. This number is put at the beginning of the sorted list that we are constructing. We proceed inductively with the rest of the list and eventually end up with a sorted list. How many comparisons do we need with this approach? In the first traversal of the list, we always need to compare the currently smallest number with the next number on the list, for a total of $n - 1$ comparisons of two numbers. More generally, during the j-th traversal of the list we need to perform $n - j$ comparisons. Overall, this would result in $\sum_{j=1}^{n-1}(n - j) = \sum_{k=1}^{n-1} k = n(n - 1)/2$ comparisons.

But we can do much better, with a procedure called *merge sort*. The idea of the procedure is that instead of sorting the entire list $a_1, \ldots, a_n$, we break the list into two approximately equal lists, one consisting of the first $\lceil n/2 \rceil$ elements of the list, the other one of the remaining $\lfloor n/2 \rfloor$ elements, we recursively call the same method on these shorter lists, and then combine the two sorted lists into a single sorted list. Of course, we need to explain how to combine two sorted lists into a single sorted list. But let us first illustrate on a concrete example how the calculation can be conveniently organized into a tree structure.

Example 9.10. Consider the following list of 13 integers: 3, 19, 6, 8, 12, 11, 10, 22, 4, 1, 7, 9, 18. The list is broken into smaller and smaller lists, until we obtain lists of size 1, which are trivial to sort, see Fig. 9.9.   ◇

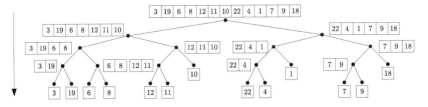

FIGURE 9.9   Simplifying the problem of sorting a list of integers.

Let us now explain how to combine two sorted lists into a single sorted list. Suppose that we have two sorted lists of integers, $b_1, \ldots, b_k$ and $c_1, \ldots, c_\ell$. We create a new list L which is initially empty but will eventually become a single sorted list combining the two given sorted lists. We introduce a pair of indices i and j, each responsible for traversing one of the two initial lists. We traverse the two lists in parallel, always keeping track of the next element for each list that is the candidate for extending the current partially built list L. We initialize $i = 1$ and $j = 1$. If we have not yet exhausted any of the two lists, that is, if $i \leq k$ and $j \leq \ell$, then we compare the current elements b_i and c_j. If $b_i \leq c_j$, we append b_i at the end of L and increase i by 1. Otherwise, we append c_j at the end of L and increase j by 1. Sooner or later one of the two original lists will be completely traversed, at which point we add the remaining elements of the other list at the end of L. More precisely, if $i \leq k$ and $j > \ell$, we append b_i at the end of L and increase i by 1, and if $i > k$ and $j \leq \ell$, we append c_j at the end of L and increase j by 1. The algorithm stops when $i > k$ and $j > \ell$, at which point L is a sorted list of integers that appeared in the combined list $b_1, \ldots, b_k, c_1, \ldots, c_\ell$. Since at every step exactly one of the indices i and j is increased by one, we conclude that the total number of comparisons performed is bounded by the sum $k + \ell$ of the lengths of the two lists.

Example 9.11. Consider the following two lists: $(b_1, \ldots, b_7) = (3, 6, 8, 10, 11, 12, 19)$ and $(c_1, \ldots, c_6) = (1, 4, 7, 9, 18, 22)$. The steps of the merging algorithm are as follows:

- Initialization: $L = ()$, $i = j = 1$.
- $c_1 = 1 < b_1 = 3$, we append c_1 at the end of L to obtain $L = (c_1) = (1)$; we increase the value of j to 2.
- $b_1 = 3 \leq c_2 = 4$, we append b_1 at the end of L to obtain $L = (1, 3)$; we increase the value of i to 2.
- The next values appended at the end of L are, in order: $c_2 = 4$, $b_2 = 6$, $c_3 = 7$, $b_3 = 8$, $c_4 = 9$, $b_4 = 10$, $b_5 = 11$, $b_6 = 12$, $c_5 = 18$, and $b_7 = 19$.
- At this point we have $i = 8$, meaning that we have used up all the elements from the first list. We append the remaining elements from the other list (in this case only one, $c_6 = 22$) at the end of L, to obtain final the merged list $L = (1, 3, 4, 6, 7, 8, 9, 10, 11, 12, 18, 19, 22)$. $\diamond$

Now that we have explained how to merge two sorted lists, the algorithm is clear: we traverse the tree that we constructed in the first step, when breaking down the problem to simpler subproblems, in the opposite direction, from smaller to larger problems. At each step, we merge two sorted lists using the procedure we described above. Once the procedure is finished, we will have sorted the initial list.

How many comparisons do we need in the worst case? If we denote by $T(n)$ the number of comparisons needed in the worst case by the merge sort procedure when sorting a list of n integers, then the following holds: $T(1) = 0$ and for all $n > 1$, we have $T(n) \leq T(\lceil n/2 \rceil) + T(\lfloor n/2 \rfloor) + n$. Using induction on k, it can be shown that for all $k \geq 0$, we have $T(2^k) \leq k \cdot 2^k$ (see Exercise 2 at the end of the section). Thus, if we denote by k the smallest integer such that $n \leq 2^k$, then $2^{k-1} < n$, or equivalently $k < \log n + 1$ and $2^k < 2n$, and we infer that $T(n) \leq T(2^k) \leq k \cdot 2^k < 2n \cdot (\log n + 1)$. This implies the following:

Theorem 9.12. *The merge sort algorithm sorts a list of n positive integers with fewer than $2n \cdot (\log n + 1)$ comparisons.*

This is much better than $n(n-1)/2$ comparisons. For example, for $n = 100$, we have at most 1528 comparisons instead of at most 4950, but the larger the value of n, the more dramatic the difference; in fact $2n \cdot (\log n + 1) = o(n(n-1)/2)$.

Let us also note that the above theorem holds in a much more general setting, namely for sorting objects from any set S equipped with a partial order $\preceq$ such that any two objects $a, b \in S$ are comparable (that is, at least one of $a \preceq b$ or $b \preceq a$ holds).

In conclusion, we return to our working example, Example 9.10.

Example 9.13 (continued). We now traverse the tree from Fig. 9.9 bottom up and at every step merge two sorted lists into a larger sorted list. See Fig. 9.10.

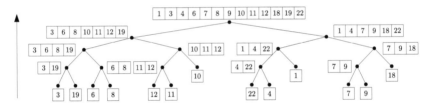

FIGURE 9.10 Merging the sorted lists iteratively from bottom to top.

As you have probably noticed, the last merging is the one from Example 9.11. ◇

Exercises

1. Sort the following list of integers using merge sort:

$$12, 22, 53, 17, 4, 9, 17, 99, 0, 8, 11, 20, 7, 18, 33, 21, 24, 5, 28, 42 \,.$$

Show all your work.

2. Let T be a function mapping positive integers to nonnegative integers such that $T(1) = 0$ and $T(2^k) \leq 2T(2^{k-1}) + 2^k$ for all $k \geq 0$. Show using induction on k that $T(2^k) \leq k \cdot 2^k$ for all $k \geq 0$.

3. Among all orderings of $\{1, 2, 3, 4\}$ find the one for which mergesort uses the largest number of comparisons.
Repeat this for $\{1, 2, \ldots, 8\}$ and $\{1, 2, \ldots, 16\}$.
Can you generalize to $\{1, 2, \ldots, 2^k\}$?

9.4 Planarity

In Section 9.2, we learned how to connect a given set of geographical locations (for example, cities or villages) by roads in a cheapest possible way. In reality, once the roads are built, the network typically contains many more roads than predicted by a solution to the Minimum Spanning Tree problem. This is because we want the network not only to be connected but to allow for overall fast travel times between different cities. On a more local scale, a similar phenomenon occurs also in networks of streets within a city.

Imagine a situation where we have five important cities that we would like to pairwise connect with direct road connections. While we are prepared to build all the corresponding $\binom{5}{2} = 10$ roads, we would like to do so without incurring additional expenses and risks related to building bridges or tunnels. Can this be achieved?

✠

If you try to solve this puzzle with paper and pencil, you will quickly realize that it is not possible to achieve the desired design of roads; at least one pair of cities must remain unconnected if bridges and tunnels are to be avoided. See Fig. 9.11 for an example, where, with the other roads in place, Zurich and Bled cannot be connected by a direct road, even if the last road were not required to be straight.

FIGURE 9.11 Five cities and roads between them.

A similar situation can occur even without requiring all possible pairs of cities to be connected. A classical mathematical puzzle known as the *three utilities problem* or sometimes *water, gas, and electricity* asks for non-crossing connections to be drawn between three houses and three utility companies. (Try it! You can place the house and companies wherever you like in the plane.)

✠

The graph in Fig. 9.11 is *planar*, which means that it can be drawn in the plane without edge crossings. The complete graph K_5 and the complete bipartite graph $K_{3,3}$ are both non-planar, there is no way to draw them in the plane even with curved edges!

When we draw a graph we represent the vertices by dots and the edges by curves.

In Fig. 9.12 we see a planar graph and several drawings of it, the leftmost drawing has crossing edges.

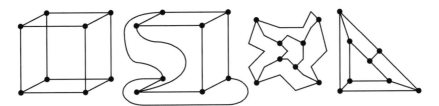

FIGURE 9.12 Drawings of the cube graph with crossing edges and without.

Cutting along the edges of a non-crossing drawing divides the plane into different regions which we call *faces*. We can describe the face set F by closed walks in the graph by imagining walking on a face close to the cuts. The triple $G = (V, E, F)$ is called a *plane graph*.

In Fig. 9.13 we indicate the faces by shadings. The inside white face is described by the walk $A, (A, b), b, (b, D), D, (D, c), c, (c, A), A$. The outside white face is described by the walk $d, (d, C), C, (C, a), a, (a, B), (B, d), d$. Note that all these walks are actually cycles.

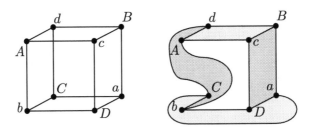

FIGURE 9.13 A drawing of the cube graph dividing the plane into 6 faces.

It seems intuitive that a simple graph with many edges, such as a complete graph on $n \geq 5$ vertices, will be impossible to draw without edge crossings. To obtain a bound on the maximal number of edges in terms of the number of vertices of a simple planar graph, we might ask the following question. How many edges does a connected plane graph G on n vertices have if all faces are triangles (i.e., cycles of length 3)?

Since G is connected, it contains a spanning tree T, and T has $n - 1$ edges. Consider a plane drawing of G and cut along the edges of T. Your piece of paper is still connected and you may consider it a $(2n - 2)$-gon, see Fig. 9.14. By a simple inductive argument you find that $k - 3$ diagonals are needed to triangulate

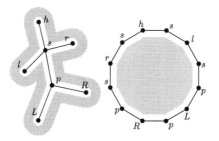

FIGURE 9.14 Turning the exterior of a tree inside out.

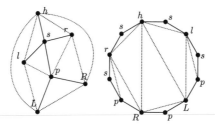

FIGURE 9.15 A triangulation.

a k-gon, see Fig. 9.15. That means G has $n - 1 + 2n - 2 - 3 = 3n - 6$ edges. For instance, you can verify this count on the graph of Fig. 9.11.

Given a plane graph $G = (V, E, F)$ let us look at the collection F of walks. It is clear that we have written every edge exactly twice. We can therefore define a *dual* graph G^* on the new vertex set F and on the same edge set E by defining two faces to be adjacent if the corresponding walks contain the edge $e \in E$. In fact, we get a plane graph $G^*(F, E, V)$ by interchanging the roles of F and V.

Fig. 9.16 shows two different planar drawings of the same planar graph, both of which have four faces: three *interior* faces and one *exterior* face (which contains the margins of your sheet of paper).

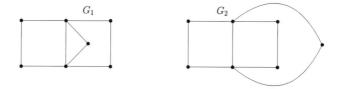

FIGURE 9.16 Two plane graphs G_1 and G_2 corresponding to the same planar graph.

To understand why these two drawings are not considered equivalent to each other, look at the faces. The graph G_1 has for example a face which is a 3-cycle, while G_2 does not. This example also shows that the dual of a simple plane graph need not be simple.

If we have a connected plane graph and apply the dual operation twice, we get back to the same graph.

Interestingly, as shown by the Swiss mathematician Leonhard Euler, while a planar graph may have many non-equivalent planar drawings, the number of faces in any such embedding is always the same; in fact, it depends only on the number of vertices and edges of the graph.

Theorem 9.14 (Euler's formula). *Let G be a connected plane graph and denote by n the number of vertices, by e the number of edges, and by f the number of faces of G. Then $n - e + f = 2$.*

Proof sketch. Consider a spanning tree T of G. If $T = G$, then Euler's formula holds as in this case $e = n - 1$ and $f = 1$. Consider the edges not in T in some order $\{e_1, e_2, \ldots\}$. Cutting along e_1 cuts the face created by T into two pieces, an interior piece, and an exterior one. Since we started with a plane graph G, e_2 is completely contained in one of the faces we have created so far, hence cuts one of the given faces into two pieces. We therefore create $e - (n - 1) + 1 = f$ faces. □

Our proof of Euler's formula is an induction proof, namely induction on e. It also provides, using the two-out-of-three theorem, a remarkable property of plane graphs: There are $f - 1$ edges not in T, they correspond to $f - 1$ edges separating two faces – or connecting two vertices of the dual graph, so they are the edges of a spanning tree (by count and connectivity) of the dual graph. A pair of complementary trees for the plane graphs of Fig. 9.16 is highlighted in Fig. 9.17.

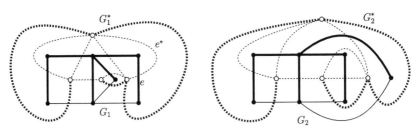

FIGURE 9.17 Complementary trees.

Using Euler's formula, we can now re-derive, as a check, our upper bound on the number of edges in a *simple* planar graph with a given number of vertices. The bound can be further improved if G is *triangle-free*, that is, it does not contain any three pairwise adjacent vertices.

Theorem 9.15. *Let G be a simple planar graph with exactly $n \geq 3$ vertices and e edges. Then $e \leq 3n - 6$. Furthermore, if G is triangle-free, then $e \leq 2n - 4$.*

Proof. Assume first that G is connected. Fix an arbitrary planar embedding of G and let us denote by f the number of faces. For a face F, the *length* of F is the number of edges in the walk bounding the face (counted with multiplicities). For example, the graph G_1 in Fig. 9.16 has faces of four different lengths, 3, 4, 5, and 6. Let us denote by S the sum of the lengths of the faces. Since $n \geq 3$ and G is simple, each face has length at least three. Thus $S \geq 3f$. On the other hand, since each edge appears on the boundary of exactly two faces (or it appears twice on the boundary of a single face), each edge contributes exactly 2 to the sum S, which implies that $S = 2e$. By Theorem 9.14, we have that $e - n + 2 = f$. Multiplying by 3, we obtain $3e - 3n + 6 = 3f \geq S = 2e$, from which the claimed inequality $e \leq 3n - 6$ follows. If G is triangle-free, then $S \geq 4f$ and hence $4e - 4n + 8 = 4f \leq S = 2e$, implying $e \leq 2n - 4$.

If G is not connected, then we can add edges to it to obtain a simple connected planar graph G' (which can be made triangle-free if G is triangle-free). The two inequalities for G' will imply the desired conclusion for G, too. $\square$

By Theorem 9.15, every simple planar graph with 5 vertices has at most 9 edges. Since the complete graph K_5 has 10 edges, we conclude that it is nonplanar. Similarly, the complete bipartite graph $K_{3,3}$ is nonplanar since it is triangle-free and has 9 edges, while every simple planar triangle-free graph with 6 vertices has at most 8 edges.

We close this section by explaining in which sense these two graphs, K_5 and $K_{3,3}$, are essentially the only reason for which a graph may fail to be planar. Let us first describe an operation that preserves planarity. Given a graph G and an edge e in G with endpoints u and v, the *subdivision* of e is the operation that replaces the edge e in G with a path of length two; formally, it deletes the edge e, adds a new vertex z, and adds two new edges, one with endpoints u and z, and one with endpoints z and v. The *subdivision* of a graph G is any graph obtained from G by repeatedly subdividing some (possibly none) of its edges. Given two graphs G and G', we say that G' *contains* G if G can be obtained from G' by a, possibly trivial, sequence of vertex and edge deletions.

Given two graphs G and G' such that G' is a subdivision of G, it is not difficult to check that G is planar if and only if G' is planar. In particular, all subdivisions of K_5 and $K_{3,3}$ are nonplanar. Furthermore, when a vertex or an edge is deleted from a plane graph, the resulting graph is still a plane graph. This means that each graph contained in a planar graph is planar. We conclude that whenever a graph contains a subdivision of K_5 or $K_{3,3}$, we can be sure that it is nonplanar. As shown by the Polish mathematician Kazimierz Kuratowski in 1930, the absence of subdivisions of K_5 or $K_{3,3}$ is not only a necessary condition for planarity but also a sufficient one!

Theorem 9.16 (Kuratowski's theorem). *A graph is planar if and only if it does not contain a subdivision of K_5 or $K_{3,3}$.*

What this means is that there is always a good way to certify the planarity or nonplanarity of a given graph. If the graph is planar, we can certify this by drawing it in the plane without edge crossings. If the graph is nonplanar, we can certify this by exhibiting an appropriate subdivision of K_5 or $K_{3,3}$ contained in the graph.

Exercises

1. Let G be the simple graph with vertex set $\{1, 2, 3, \ldots, 9\}$ in which two distinct vertices are adjacent if and only if they are coprime. The graph H is defined similarly, except that two distinct vertices are adjacent if and only if they are not coprime.
(a) Show that G is not planar by showing that it has too many edges / that it contains a $K_{3,3}$ / that it contains a K_5.
(b) Show that the graph H is planar by drawing a planar embedding of it.
2. Derive the bound for the maximal number of edges in a bipartite simple planar graph.
3. Draw a plane embedding of K_4 and its dual graph. Verify that for each spanning tree the complementary edges form the edges of a dual spanning tree.

9.5   Eulerian graphs

In this section and the next one, we will be looking at two further problems on graphs, with obvious interpretations in practical settings where the graph represents a network of streets within a city or a network of roads between cities.

The first problem is the following: Given a graph G, can all the edges of G be traversed so that no edge repeats? Graphs for which the answer is affirmative can be described by means of a simple criterion discovered in 1736 by Euler. The paper in which he solved the special case of the problem known as the *Seven Bridges of Königsberg* (see Exercise 1 at the end of the section) is now widely considered as the first paper of graph theory.

The problem asks about the existence of an *Eulerian trail* in a graph, a traversal of all the edges such that no edge repeats. An Eulerian trail starting and ending in the same vertex is said to be an *Eulerian circuit*. A more formal definition is as follows.

Definition 9.17. Given a graph G, a *trail* in G is a walk

$$v_0, e_1, v_1, e_2, v_2, \ldots, v_{k-1}, e_k, v_k$$

without repeated edges. An *Eulerian trail* in G is a trail containing all the edges of G. An *Eulerian circuit* is an Eulerian trail with $v_0 = v_k$. A graph is said to be *Eulerian* if it admits an Eulerian circuit. ♠

For example, the graph depicted in the left part of Fig. 9.18 is Eulerian, as evidenced by the traversal of the edges depicted in the right part of the figure. Note that vertex repetitions are perfectly ok.

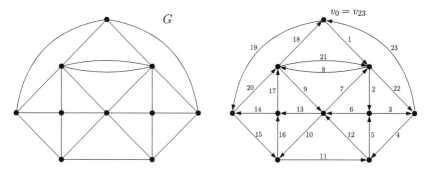

FIGURE 9.18   A graph with 23 edges and an Eulerian circuit in it.

An *isolated vertex* in a graph is a vertex that is not the endpoint of any edge. If a graph has an isolated vertex v, then deleting v does not affect the existence of Eulerian trails or circuits. Therefore, when addressing this problem we may without loss of generality restrict our attention to graphs without isolated vertices. Another easy observation is that if a graph has an Eulerian trail, then any two of its edges must be connected by a path; that is, all the edges of the graph must belong to the same connected component. (Do you see why?)

It is not difficult to construct examples of connected graphs that admit an Eulerian trail but not an Eulerian circuit, or connected graphs that do not admit an Eulerian trail.

Example 9.18. In Fig. 9.19, two small connected graphs G_1 and G_2 are depicted.

FIGURE 9.19   Two graphs G_1 and G_2.

The graph G_1 admits an Eulerian trail but not an Eulerian circuit. Traversing the only edge in either direction yields an Eulerian trail. Since we cannot return to the starting vertex without traversing the same edge again, the graph does not admit an Eulerian circuit.

The graph G_2 does not admit an Eulerian trail. Do you see why?

Every time an Eulerian trail in a graph visits a vertex, it either has to stop at that vertex or it has to leave the vertex on a yet unvisited edge. Therefore, the edges around every vertex, except possibly for the starting and the ending vertex, can be grouped into disjoint pairs, one for each visit of the vertex. Since in an Eulerian trail all the edges are traversed, we can conclude that all vertices of the graph, except possibly two, must have an even degree. Going back to our example graph G_2, note that all the four vertices of the graph have an odd degree. It follows that G_2 does not have any Eulerian trails. $\Diamond$

As observed above, a necessary condition for a connected graph to have an Eulerian trail is that it has at most two vertices of odd degree. Since every graph has an even number of vertices of odd degree (see Exercise 2 at the end of the section), this is equivalent to the condition that the number of odd-degree vertices is either 0 or 2. Euler showed that this obvious necessary condition for the existence of an Eulerian trail is also sufficient. Furthermore, the absence of odd-degree vertices leads to a criterion for the existence of an Eulerian circuit.

Theorem 9.19. *Let G be a graph without isolated vertices. Then G has an Eulerian trail if and only if G is connected and the number of odd-degree vertices is either 0 or 2. Furthermore, G has an Eulerian circuit if and only if G is connected and all its vertices have an even degree.*

Proof. Assume first that G has an Eulerian trail $v_0, e_1, v_1, e_2, v_2, \ldots, v_{m-1}, e_m, v_m$. Since G has no isolated vertices, in order to show that G is connected, it suffices to show that any two edges e_i and e_j, $i < j$, are connected by a walk in G. Such a walk can be found by following the part of the Eulerian trail from e_i to e_j. Thus, G is connected. We have already presented in Example 9.18 a pairing argument showing that all the vertices of G except possibly v_0 and v_m have even degree. If G has an Eulerian circuit, with $v_0 = v_m$, then the first and the last edge in the circuit have v_0 as a common endpoint, and we can also pair these two edges up. In this case, all the vertices of G, including v_0, must have an even degree.

Next, let us show that the conditions are sufficient. Assume first that G is connected and all its vertices have even degree. Pick a vertex v and start traversing edges in v as long as possible, subject to the condition that no edge is ever repeated. Since the graph is finite, this process eventually stops and computes some trail in G starting at v. Let us denote by w the final vertex of this trail, that is, the vertex in which the process stopped. If $w \neq v$, then an odd number of edges around w was used: one for the last visit of w (after which it was not possible to leave w), and two for each previous visit. However, since w has an even degree, there is an edge at w that was not yet traversed, hence our traversal could not stop at w. This contradiction shows that the process stopped at v, that is, we obtained an Eulerian circuit of the graph formed by the edges traversed so far.

Note that we used an even number of edges around each vertex. Since all the degrees are even, this means that the number of remaining edges around each vertex is also even. If we already used up all the edges of the graph, then we have constructed an Eulerian circuit. So we may assume that this is not the case. Since the graph is connected, there exists a vertex z that is an endpoint of a traversed edge as well as an endpoint of an edge that has not yet been traversed. We now repeat the same procedure starting at the vertex z and using only the edges not used so far. Using the same arguments as above, we obtain another trail starting and ending at the vertex z. Combining the two trails by traversing the first one until the vertex z, then traversing the second one, and finally continuing along the first one until the end, results in an Eulerian circuit of the graph formed by the edges traversed so far. Repeating the procedure, we eventually obtain an Eulerian circuit in the entire graph G.

It remains to analyze the case when G is connected and has exactly two vertices of odd degree, say u and v. Adding to G an edge e with endpoints u and v we obtain a connected graph G' in which all vertices have an even degree. Therefore, G' has an Eulerian circuit. Fix an Eulerian circuit $v_0, e_1, v_1, e_2, v_2, \ldots, v_{m-1}, e_m, v_m$ in G'. Since $v_0 = v_m$, we may assume, by a circular shifting of the indices if necessary, that the edge e appears at the end of the circuit, that is, $e_m = e$. But then $v_0, e_1, v_1, e_2, v_2, \ldots, v_{m-2}, e_{m-1}, v_{m-1}$ is an Eulerian trail in G. □

The above proof suggests a procedure to construct an Eulerian circuit in a connected graph in which all vertices have an even degree. It can be shown that it is always possible to select the next edge so that the edges not yet used form a connected graph. Following this rule, we end up with an Eulerian circuit.

Exercises

1. The city of Königsberg in Prussia (now Kaliningrad, Russia) was set on both sides of the Pregel River. There were two large islands and two mainland portions of the city, which were connected to each other by seven bridges. The problem of the *Seven Bridges of Königsberg* was to devise a walk through the city that would cross each of those bridges exactly once. Euler showed that such a walk exists if and only if the graph depicted in Fig. 9.20 contains an Eulerian trail. Assuming this, explain why the desired walk through the city does not exist.

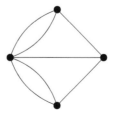

FIGURE 9.20 The graph of the seven bridges of Königsberg.

2. Show that in any graph G, the number of odd-degree vertices is even.

3. Fig. 9.21 shows three graphs. For each of them determine if the graph contains an Eulerian trail and whether it contains an Eulerian circuit. If an Eulerian trail exists, find one.

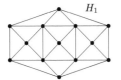 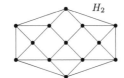 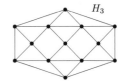

FIGURE 9.21   Three graphs.

9.6   Hamiltonian graphs

In the previous section we considered the problem of traversing all the edges of a given graph so that no edge repeats. Following Euler, we gave a necessary and sufficient condition for a graph to possess such a walk. In this section we turn to a similar question, but for vertices: Given a graph G, can all the vertices of G be traversed so that no vertex repeats? A walk with such a property is called a *Hamiltonian path*.

In the case of Eulerian trails, we allowed for the last vertex of the trail to coincide with the first one. It is natural to allow this also for walks traversing vertices – just think of a tourist who wants to visit all the main attractions in the city before returning back to the hotel room. Such a walk traversing all the vertices, without repetitions except that the first and the last vertex coincide, is called a *Hamiltonian cycle*. Hamiltonian paths and cycles are named after William Rowan Hamilton, a 19th century Irish mathematician, astronomer, and physicist who invented a mathematical game which involves finding a Hamiltonian cycle in the edge graph of the dodecahedron (see Fig. 2.1 on p. 43).

Definition 9.20. Given a graph G, a *Hamiltonian path* in G is a path containing all the vertices of G. A *Hamiltonian cycle* in G is a cycle containing all the vertices of G. A graph is said to be *traceable* if it possesses a Hamiltonian path and *Hamiltonian* if it possesses a Hamiltonian cycle.   ♠

If we traverse any Hamiltonian cycle in a graph G and stop the traversal just before returning to the initial vertex, we obtain a Hamiltonian path. Therefore, every Hamiltonian graph is traceable. The converse is not true. (Do you see why?)

Example 9.21. Consider the three connected graphs G_1, G_2, and G_3 depicted in Fig. 9.22. The graph G_1 is Hamiltonian (and therefore traceable); the edges traversed by one particular Hamiltonian cycle in G_1 are depicted thick. The graph G_2 is traceable; a Hamiltonian path in G_2 is depicted with thick edges. However, the graph G_2 is not Hamiltonian. The graph G_3 is not traceable (and therefore not Hamiltonian).

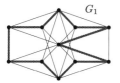 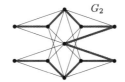 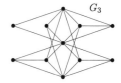

FIGURE 9.22   Three graphs that behave differently with respect to Hamiltonian paths and cycles.

Do you see why the graph G_2 is not Hamiltonian and why the graph G_3 is not traceable?

☩

Deleting any $k \geq 1$ vertices from a cycle results in a graph with at most k connected components. Consequently, if we delete any k vertices from a Hamiltonian graph, we obtain a graph with at most k connected components. Similarly, deleting any $k \geq 0$ vertices from a path results in a graph with at most $k + 1$ connected components. Hence, if we delete any k vertices from a traceable graph, the resulting graph has at most $k + 1$ connected components. Let us now apply these considerations to the graphs G_2 ad G_3. If we remove from the graph G_2 the middle vertex, the top vertex, and the bottom vertex, we obtain a graph with 4 connected components. We conclude that G_2 is not Hamiltonian. Similarly, if we remove the middle, top, and the bottom vertices from the graph G_3, we obtain a graph with 5 connected components, and we conclude that G_3 is not traceable.   ◇

Note that if a graph G has a loop, then G is traceable if and only if the graph obtained from G by deleting a loop is traceable, and a similar argument holds if G contains a pair of edges with the same endpoints. Furthermore, except for graphs with one or two vertices, similar simplifications on the graph can be done when studying the property of being Hamiltonian. Therefore, when studying Hamiltonian paths and cycles, it suffices to consider simple graphs.

Although perhaps surprising at first, you may have guessed from the above examples that the problem of determining the existence of Hamiltonian paths and cycles is much more difficult than for Eulerian trails. No simple necessary and sufficient condition for a graph to be Hamiltonian or traceable is known, and the problems of deciding if a given graph possesses a Hamiltonian path or a Hamiltonian cycle are NP-complete – just like the problem we talked about in Section 4.7, of determining if a given expression in conjunctive normal form admits a Boolean assignment to the variables which makes the expression TRUE. This means that no fast solution methods for the Hamiltonicity and traceability problems are known. These problems become difficult to solve already for moderately sized graphs.

Some necessary conditions are known, as well as some sufficient ones. We have already described in the above example the following necessary conditions.

Theorem 9.22. *If G is a traceable graph, then for every set $S \subseteq V(G)$ the graph $G - S$ has at most $|S| + 1$ connected components. Furthermore, if G is Hamiltonian, then for every nonempty set $S \subseteq V(G)$ the graph $G - S$ has at most $|S|$ connected components.*

We would expect a simple graph to have a better chance to be Hamiltonian if all the vertices are connected by edges to many other vertices. This is indeed the case, as shown by the following sufficient condition for Hamiltonicity due to Dirac.

Theorem 9.23. *Let G be a simple graph with $n \geq 3$ vertices such that each vertex has degree at least $n/2$. Then G is Hamiltonian.*

Proof. The proof uses induction in a particularly clever way. Fix a positive integer $n \geq 3$ and let us denote by X_n the set of all n-vertex graphs in which each vertex has a degree at least $n/2$. Associate to each graph $G \in X_n$ the number $\overline{e}(G)$ of *non-edges* in G, that is, the value of $\binom{n}{2} - |E(G)|$. Then $\overline{e}(G) \geq 0$. To show that each graph $G \in X_n$ is Hamiltonian, we use induction on the value of $\overline{e}(G)$.

Base case: $\overline{e}(G) = 0$, that is, the graph $G \in X_n$ contains all possible edges between two distinct vertices. Since $n \geq 3$, visiting all the vertices of G in some order and returning back to the first vertex yields a Hamiltonian cycle.

Inductive step: Let $k \geq 0$ be given and assume that every graph $H \in X_n$ with $\overline{e}(H) = k$ is Hamiltonian. Consider a graph $G \in X_n$ with $\overline{e}(G) = k + 1$. We need to show that G is Hamiltonian. Suppose for a contradiction that G is not Hamiltonian. Since $\overline{e}(G) > 0$, there exist two nonadjacent vertices in G, say u and v. The graph H obtained from G by adding to it the edge $\{u, v\}$ belongs to X_n and satisfies $\overline{e}(H) = \overline{e}(G) - 1 = k$. Therefore, by the induction hypothesis, H is Hamiltonian. Fix a Hamiltonian cycle C in H. Since G is not Hamiltonian, the cycle C must use the added edge $\{u, v\}$. We may assume that the cycle C starts at u but not along the edge $\{u, v\}$ and hence there is an ordering $v_1, \ldots, v_n$ of the vertices of G such that $u = v_1$, $v = v_n$, and any two vertices that are consecutive in the ordering are adjacent in G. (In particular, G is traceable.)

Let A denote the set of neighbors of u in G and let $B = \{v_{i+1} \mid 1 \leq i \leq n - 1$ and vertices v and v_i are adjacent in $G\}$. Since each vertex in G has degree at least $n/2$, we infer that $|A| \geq n/2$ and, similarly, $|B| \geq n/2$. Furthermore, since $u = v_1$ belongs to neither A nor B, we have $A \cup B \subseteq \{v_2, \ldots, v_n\}$ and thus $|A \cup B| \leq n - 1$. It follows that the sets A and B cannot be disjoint, as that would imply $|A \cup B| = |A| + |B| \geq n$. Using an arbitrary vertex $v_j \in A \cap B$, we can now construct a Hamiltonian cycle in G, as follows. Starting at u, go to v_j (note that this is possible since $v_j \in A$ is adjacent to u), then continue via v_{j+1} all the way up to $v_n = v$, then go to v_{j-1} (note that this is possible since $v_j \in B$ implies that v_{j-1} is adjacent in G to v), and go back all the way down to $v_1 = u$. We conclude that G is Hamiltonian, contradicting our assumption that G is not Hamiltonian. This completes the inductive step and with it the proof of the theorem. $\square$

Despite the difficulty of the Hamiltonicity problem, variants of the problem have many important practical applications. Of particular relevance is the *Traveling Salesman Problem*, which is defined similarly as the *Minimum Spanning Tree* but much more difficult. The traveling salesman would like to visit a number of cities before returning back to his home town. The distance between any two cities is known. What is the least distance that the salesman must travel in order to visit all the cities?

Exercises

1. Describe all traceable trees. Are there any Hamiltonian trees?
2. For each of the graphs in Fig. 9.23, determine whether it is traceable and whether it is Hamiltonian. Justify your answer.

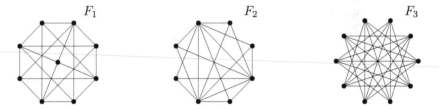

F_1 F_2 F_3

FIGURE 9.23   Three small graphs.

3. Find an example of a connected graph with 6 vertices and 8 edges that is:
   (a) Eulerian but not Hamiltonian,
   (b) Hamiltonian but not Eulerian,
   (c) Eulerian and Hamiltonian,
   (d) neither Eulerian nor Hamiltonian.
   Can you always find a simple graph with the stated properties?

9.7   Case study: Fáry's theorem

Say we have a connected planar graph and we want to draw it so that all edges are straight line segments that do not cross (except at the endpoints). Well, if there is a loop, this is impossible. Also parallel edges make the task impossible. But is it possible if the graph is simple?

   In a simple graph there are no cycles with fewer than three edges, so the smallest face one can have is a triangle, that is, a 3-cycle. Not all triangles need to be faces. If a simple plane graph has faces of larger size, we can triangulate the drawing, adding $k - 3$ new edges to each face with k edges. A look at Fig. 9.24 shows a triangulation in the upper left corner drawn with curved line segments, and re-drawn with straight line segments in the lower right. Check carefully that every face is a triangle. There is also a triangle, namely 156, which is not a face and contains the vertex 2 in its interior. If all faces are triangles, we cannot add any more edges and we speak of an edge maximal planar graph or a triangulation. István Fáry showed in 1948 that if a finite simple graph G can be represented in the plane at all, then it can be represented in the plane with straight segments as edges. We now sketch a proof of Fáry's theorem.

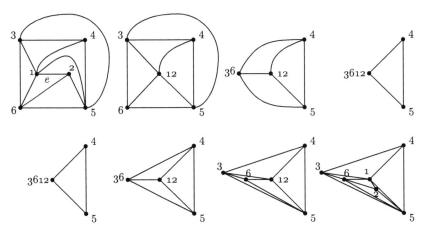

FIGURE 9.24   Fáry's induction step applied repeatedly.

Theorem 9.24. *An edge maximal simple planar graph can be drawn in the plane with non-crossing straight line segments as edges.*

Proof. Consider an edge maximal simple planar graph G. We want to show by induction on the number n of vertices that G can be drawn with straight line segments. This is clear for $n = 3$. Note that in this base case, $G = K_3$, we have three vertices, 3 edges, and two faces. The triangle cuts the plane into two pieces, an interior and an exterior.

For $n > 3$ we want to first prove by strong induction that there always exists a vertex v in the interior of the outside triangle and an edge e with endpoint v such that e is contained in exactly two triangles. For $n = 4$, the graph is a K_4 and the statement holds trivially, since any edge belongs to exactly two trian-gles. Assume now that $n > 4$ and that the statement holds for all triangulations with fewer than n vertices. Let G be a triangulation with n vertices and let v be any vertex in the interior of the outside triangle. Take any edge e with end-point v. If the edge e is contained in exactly two triangles, we are done. So we may assume that the edge e is contained in two facial triangles and also in a third triangle T. The triangle T contains exactly one of the two triangular faces separated by e. So T contains a vertex in the interior. Let H be the subgraph of G consisting of the triangle T and all the vertices and edges drawn in the interior of T. Then H is an edge maximal simple planar graph with at least 4 vertices with fewer vertices than G. By the induction hypothesis, there exists a vertex w of H in the interior of the triangle T and an edge $f \in E(H)$ with endpoint w such that f is contained in exactly two triangles in H. Due to the planarity of the drawing, this vertex and edge satisfy the same properties also with respect to the graph G.

Now let G be any edge maximal simple planar graph on n vertices, and consider an edge $e = \{a, b\}$ which separates the facial triangles (a, b, v_1), and (a, b, v_2) and is contained in no other triangle. Identifying the two endpoints a and b into a vertex v creates parallel edges $\{a, v_i\}, \{b, v_i\}, i = 1, 2,$ but no other parallel edges. Replacing each of these pairs of parallel edges by a single edge $\{v, v_i\}$ we obtain a simple graph on $n - 1$ vertices that is a planar triangulation and can, by induction hypothesis, be drawn with straight lines. In such a straight line drawing consider the edges emanating from v. One of the two angles formed by edges $\{v, v_1\}$ and $\{v, v_2\}$ contains all edges inherited from G which had endpoint a in G, the other all the ones which had endpoint b in G. So we can split v into two vertices a and b a short distance apart, draw in the straight line segment from a to b, and split the line segments $\{v, v_i\}$ into two line segments from a, resp. b, to v_i. This yields a straight line drawing of G. $\qquad\square$

Fig. 9.24 illustrates how the proof can be turned into an algorithm. The induction step applied repeatedly reduces the triangulation to a triangle – and then we may work backwards, reconstructing the graph step by step, but with straight line edges.

Could we have started the process by identifying vertices 1 and 5? The answer here is no, since the edge with endpoints 1 and 5 is contained in more than two triangles, and identifying 1 and 5 would produce three sets of parallel edges. However, you may want to convince yourself that e is not the only choice for starting the procedure. Also, the straight line drawing produced is not necessarily optically pleasing; you are encouraged to turn Fáry's proof into a nice drawing algorithm.

Note that the proof of Theorem 9.24 involves Euclidean geometry and is not completely discrete.

We proved Fáry's theorem for triangulations, but if we want a straight line embedding of a plane graph that is not edge maximal, we can first triangulate all faces that are larger than triangles, find a straight line embedding of that triangulation and then erase the extra edges to obtain a straight line drawing of our original graph. Triangles are always convex, but k-gons for $k > 3$ need not be, so after deleting edges from a triangulation, the resulting faces need not be convex. However, Sherman K. Stein, in 1951, proved that for a plane graph whose faces are polygons without repeated vertices or edges, and the intersection of two faces is connected, there always exists an embedding such that all faces are convex. The key lemma used in the induction proof on the number of faces is claiming the existence of two neighboring faces A and B such that for all other faces C the set $C \cap (A \cup B)$ is connected. The proof of this lemma is essentially the same as our argument about the existence of an edge e contained in exactly two triangles. From Stein's result Fáry's theorem follows. Convexity is a geometric property and the faces here are considered closed and simply connected (in the topological sense) subsets of the Euclidean plane. We encourage you to work through Fáry's and Stein's original papers and determine which of the two results is more discrete.

9.8 Case study: Towers of Hanoi

The first thing we noted about the Towers of Hanoi with n disks was that we could encode each state with a string of n digits from $\{0, 1, 2\}$. So 2221010 has the largest three disks on post 2 and the remaining alternating from post 1 and 0.

If we say that two states are related if there is a legal move between them, then we have a self-relation which is anti-reflexive and symmetric, in other words, the relation diagram can be expressed as a simple graph. What can we say about this graph on 3^n vertices? Our first reaction might be that the graph with only 7 disks has over two thousand vertices, so only a madman would ever want to draw or study it. But we already have been studying it. For instance, we have shown that there is a sequence of $2^7 - 1$ legal moves from 2222222 to 0000000. In the language of graph theory, there is a path of length $2^7 - 1$ joining vertex 2222222 to 0000000 in the Towers of Hanoi graph. We also showed that there was a sequence of moves from any configuration to 0000000, and we can express this graph theoretically by saying that the Towers of Hanoi graph is connected.

Maybe it is a tree? At most states there are exactly three moves. The littlest disk can move to either of the two posts it is not already on. Between those two other posts, if nonempty, the smallest top disk can move, but it cannot move to the post of the very smallest disk. That third move only fails to be possible if all the disks are stacked under the littlest disk, i.e., for the three states 0000000, 1111111, and 2222222 when $n = 7$. So in the Towers of Hanoi graph, $3^n - 3$ vertices have degree 3 and just 3 vertices have degree 2. Counting the edges in the usual way gives $|E| = [(3^n - 3)3 + 2(3)]/2 = (3^n)3/2 - 1.5 > 3^n - 2$, too many to be a tree even for n very small. But maybe you already saw an easier way to explain why the graph is not acyclic? Still the graph with 7 disks has over 3000 edges. Very big.

Let's try to build it anyway, recursively. Take the state diagram for n disks, and imagine a larger $n + 1$st disk underneath everything, but staying put on one post. As it sits there, all the other disks n are free to move, just as before. There are three possibilities for the lazy big disk, and each gives rise to a part of the state diagram for the $n + 1$ disk graph, each is an isomorphic copy of the n-disk graph, but with and additional number, not changing, appended to on the left of each vertex label. We have almost drawn the whole thing! All that is missing is that, instead of 3 vertices of degree 2, we have 9. What are they? For the biggest disk on post 0 they are $00 \cdots 0$, $01 \cdots 1$, and $02 \cdots 2$. The other six are $10 \cdots 0$, $11 \cdots 1$, $12 \cdots 2$. $20 \cdots 0$, $21 \cdots 1$, and $22 \cdots 2$. Of these $00 \cdots 0$, $11 \cdots 1$, and $22 \cdots 2$ have degree 2 in the $n + 1$ disk graph. The other six are connected with the following edges: $(10 \cdots 0, 20 \cdots 0)$, $(01 \cdots 1, 21 \cdots 1)$, and $(02 \cdots 2, 12 \cdots 2)$.

So, recursively, we just take three copies of the n-disk graph and add three edges between them, only joining some of the special, degree 2 vertices. What do we start with? Just a triangle! I won't draw it for you, in fact, in this section you have to do *all* the drawing. Try. The four disk graph takes four minutes.

So now you have a beautiful picture of the Towers of Hanoi graph. It's not beautiful? Did you notice that the graph is planar? Before you try again, we'll prove that outrageous claim. We'll do it by induction and as often happens, it is easier to prove a somewhat stronger statement. *The Towers of Hanoi graph for n disks has a planar drawing in which all the degree 2 vertices are on the exterior face.* For the base case, the graph is a triangle, so the claim is true, every vertex is on the exterior face. Now let $n \geq 0$ be given and represent the graphs on a piece of paper by three blobs and label the 9 special vertices wherever you like on the three blob boundaries. Can you join them correctly by edges in just the exterior? You might need squiggly edges, but, unlike the similar but impossible water-gas-electric problem, this should only take you a few seconds. Now, are all the remaining special vertices, the degree two ones, on the exterior face? No? No problem. The blobs are only attached at two points, so they be redrawn twisted upside down. Done. Now that you know that the graph is planar, try to make a better picture.

Here is another fact you can prove by induction, just add it to the inductive hypothesis above and it will mostly slide right through: *There is a planar drawing with the optimal path from all 0's to all 2's laying entirely along the exterior face.* Here is another, *there a Hamiltonian path joining state 0000000 to 2222222.* That corresponds to a sequence of moves starting with 0000000, passing exactly once through every possible state, and ending 2222222. Don't you wonder what it looks like?

Too bad that there isn't a Hamiltonian cycle. But maybe there is. Why don't you recycle your simplified water-gas-electric blob diagram and see if that can help you prove that there is one. All these results you can describe and prove without the graph, or the pictures, but why would you? And while it may be true that only a madman would actually draw a graph on two thousand vertices and three thousand edges, a mathematician will certainly try to imagine it, and a computer scientist will certainly try to program a computer to render it.

9.9 Case study: Anchuria

An old Anchurian[1] custom lends itself very well to a two-person adventure, or a class experiment.

Cut six pieces of twine, each six inches in length, align them and grab them with your fist. Have your partner tie 3 knots on top and 3 knots on the bottom, see Fig. 9.25.

FIGURE 9.25 A two-person adventure.

Let go. What is the result? The Anchurian custom says that if the 6 strings form a ring, you may marry your partner.

Interpreting the 6 strings as edges and the knots as vertices, we see that the result of the experiment is a graph G on 6 vertices and 6 edges. The graph is bipartite, since every edge goes from top to bottom. It is also 2-regular, i.e., every vertex is of degree 2 (since the strings were tied two by two). How can we count the number of 2-regular bipartite graphs on six vertices? Let us first assume that we are counting graphs whose edges are labeled $1, \ldots, 6$. For tying the top knots, by the multiplication principle, there are $\binom{6}{2}\binom{4}{2}\binom{2}{2}$ possibilities, and we can divide this number by 6, since we are not interested in which order the knots were tied, only in the resulting three knots. The same holds for the bottom three knots, so there are 15^2 possibilities to tie the knots. After the bottom knots are tied, there are 15 ways to tie the top knots, we list them in Fig. 9.26. You see that 8 out of the 15 outcomes are connected, 6 are consisting of a 4-cycle and a 2-cycle, and one consists of three 2-cycles. Let us call two graphs isomorphic if there is a bijection between their vertex sets preserving adjacency. Graph isomorphism induces an equivalence relation on the set of graphs produced by the experiment and we see that there are three isomorphism classes.

We could consider the experiment having three different outcomes: three 2-cycles, a 2-cycle and a 4-cycle, or a 6-cycle. We can put our counting techniques to work to answer the question about likelihood. What is the probability to obtain a 6-cycle? It is $\frac{8}{15} > \frac{1}{2}$.

Now we want to discretely generalize the experiment. Instead of 6 strings we want to consider $2n$ strings (why do we want an even number of strings?). It

---

[1] Anchuria is a fictional country invented by O. Henry. A. Engel, well known for his mathematical problem collections and pedagogy, uses this setting to describe several interesting problems.

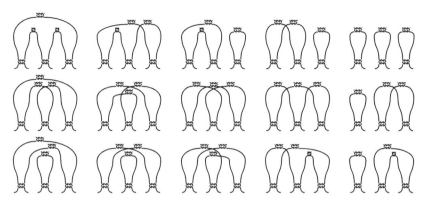

FIGURE 9.26 Ways to tie the knots.

is quite easy to describe the isomorphism classes of possible outcomes: bipartite graphs on $2n$ vertices which are the disjoint union of (non-empty) even cycles.

But now, what is the probability of obtaining a connected graph, i.e., a cycle of length $2n$? How does this probability behave as n gets large?

As before, the number of tying knots on top and bottom is $(\frac{1}{n!} \prod_{k=0}^{n} \binom{2n-2k}{2})^2$.

The number of ways to get a $2n$-cycle is $\frac{1}{n!} \prod_{k=0}^{n-1} \binom{2n-2k}{2} \prod_{k=1}^{n-1} 2k$.

So the probability of getting a $2n$-cycle is $\frac{(2n-2)(2n-4)...2}{(2n-1)(2n-3)...1}$, which goes to zero as n tends to infinity (remember Calculus?).

We could generalize also by taking three strings at a time and tie them into a knot, requiring $3n$ strings for the experiment. Try it! Is connectivity of the outcome likely if n gets large? How about tying k strings at a time? What the experiment is creating is a random regular bipartite graph.

Something else occurs when you are doing this experiment. Sometimes the cycles are linked. We are leaving the discrete world.

9.10 Summary exercises

You should have learned about:

- Graphs and subgraphs
- Paths, cycles, trees, complete graphs, bipartite graphs
- Connectivity and connected components of a graph
- Trees, their characterization, and uses in searching and sorting
- Spanning trees and how to find them
- Eulerian and Hamiltonian graphs
- Planar graphs and their duals

1. Given two simple graphs, $G_1 = (V_1, E_1)$ and $G_2 = (V_2, E_2)$.
   If G_2 is a subgraph of G_1, then $V_2 \subseteq V_1$ and $E_2 \subseteq E_1$.
   Show that the converse of the implication need not hold.

2. $G = (V, E)$ be a graph. Show that the connectedness relation, $\leadsto$, defines an equivalence relation on the vertex set.

3. Let $G(V, E)$ be a simple graph, with both V and E non-empty, $V' \subseteq V$, and $E' \subseteq E$. Give examples such that $G(V', E')$ is NOT a graph. Give examples such that $G(V', E')$ IS a graph.

4. How many connected components can a simple graph on 10 vertices and 5 (or 6 or 7) edges have at most? At least? Give examples!

5. How many connected components can a simple graph on n vertices and $\binom{k}{2}, k < n$ edges have at most?

6. Prove that a graph G is bipartite if and only if all cycles of G have even length.

7. Let G be a graph whose vertex set are bit strings of length 3 and two bit strings are adjacent if they differ in exactly one bit. Draw G. Is G simple? Is G connected? Is G planar? Is G bipartite? Is it Eulerian? Is it Hamiltonian?

8. Prove the Two-out-of-three Theorem for trees.

9. Show that the cube graph (consisting of the 8 vertices and 12 edges of the cube) is bipartite and planar. Is it edge maximal under the conditions of being simple, planar and bipartite?

10. Show that the cube graph plus one of the major diagonals is not planar.

Selected answers and solutions

Chapter 1 – Discreteness

1.2 Of course this is subjective, but Morse code seems discrete in a way that hula dancing is not.

1.3 I say frissbeeee ... to h-o-p-s-c-o-t-c-h.

2.1 Hint: If the students all answer honestly, you cannot assume independence. Why?

2.2 You can analyze the problem by considering the no-hood option separately, or taking "no hood" to be another color "invisible". So the answer is $4 \cdot 5 \cdot 2 = 40$ or $4 \cdot 4 \cdot 2 + 4 \cdot 2 = 32 + 8 = 40$.

2.3 Hint: I would say that if you don't take the entrée, you are not ordering the special. All other courses are optional, including taking the salad with no dressing.

3.1 $\binom{10}{5}9^5 + \binom{10}{6}9^5 + \binom{10}{7}9^3 + \binom{10}{8}9^2 + \binom{10}{9}9^1 + \binom{10}{10}9^0$.

3.2 $\binom{10}{5}5^{10}$.

3.3 Billiard balls are numbered and colored. We say they are *distinguishable*. So for each ball, we have to choose a compartment. These drawer compartments are also distinguishable. So for each ball, choose one of the three compartments. Fifteen choices of 3 give 3^{15}. (Not 15^3.) What if they were ping-pong balls?

4.1 The circled entries are placed symmetrically, or did you forget to count from zero?

4.2 $\binom{100}{50}/\binom{100}{49} = \frac{100!49!51!}{100!50!50!} = \frac{49!51!}{50!50!} = \frac{49!50!(51)}{49!(50)50!} = \frac{51}{50} > 1$.

4.3 Hint: You have 8 children, 4 boys and 4 girls, and want to count the number of groups of size 4, so $\binom{8}{4}$ since sex is irrelevant to the question. But what if you want to take it into account anyway?

5.1 The first two are easy. The third is a shift, notice the "binary point". The last should look familiar, and note that it is easier if you write the carries also in binary.

5.2 $2^{11} - \binom{11}{3} - \binom{11}{2} - \binom{11}{1} - \binom{11}{0}$.

5.3 First is 1 followed by 16, second has 64 zeros.

6.1 2020 in binary is 1100101000000 using either method. If you are like most people, you add commas to make it easier to read: 1,100,101,000, 000. The numbers in between the commas are 3 digit binary numbers, so 0 to 7, – aha – the digits in base 8, that's the trick! Base 8: 14500, Base 16: 1,1001,0100,0000, or 1940.

6.2 22212102 and 201324.

6.3 1776 is ↑↓↑↑ 0 ↑↓↑ 0.

Chapter 2 – Basic set theory

1.1 It is easy to find problems with each one.

1.2 It is hard to avoid problems with natural language. (∅ is well defined.)

1.3 $A = \{k \in \mathbb{Z} \mid k = 2j+1; j \in \mathbb{Z}\}$, $D = \{x \in \mathbb{R} \mid x = \sum_{i=J}^{K} d_i \cdot 10^i; J, K \in \mathbb{Z}; J \leq K; d_i \in \{5,6\}\} \cup \{x \in \mathbb{R} \mid x = \sum_{i=-\infty}^{K} d_i \cdot 10^i; K \in \mathbb{Z}; d_i \in \{5,6\}\}$.

2.1 For a) there are $2^4 = 16$ elements to be listed. For b) there are $\binom{5}{2} = 10$: $\{a, b\}, \{a, c\}, \{a, d\}, \{a, e\}, \{b, c\}, \{b, d\}, \{b, e\}, \{c, d\}, \{c, e\}$, and $\{d, e\}$.

2.2 $\mathcal{P}_3(\mathcal{P}_2(\mathcal{P}_1(\mathbb{Z})))$ consists of sets of three elements, like
$\{\{\{1\}, \{2\}\}, \{\{1\}, \{3\}\}, \{\{1\}, \{4\}\}\}$,
$\{\{\{1\}, \{2\}\}, \{\{1\}, \{3\}\}, \{\{1\}, \{5\}\}\}$, and
$\{\{\{1\}, \{2\}\}, \{\{1\}, \{3\}\}, \{\{1\}, \{6\}\}\}$.

2.3 Argue: Let $X \in \mathcal{P}_2(\{1, 2, 3, 4\})$, so $X = \{a, b\}$, with $1 \leq a, b \leq 4$. So $a, b \in \{1, 2, 3, 4, 5, 6\}$, hence $X = \{a, b\} \subseteq \{1, 2, 3, 4, 5, 6\}$, as required.

3.1 c) Sets of the form $\{(a, b), (a, c), (b, b), (b, c)\}$. There are $\binom{4}{2} = 6$ of them: $\{(0, \emptyset), (0, \pi)\}$, $\{(0, \emptyset), (\emptyset, \emptyset)\}$, $\{(0, \emptyset), (\emptyset, \pi)\}$, $\{(0, \pi), (\emptyset, \emptyset)\}$, $\{(0, \pi), (\emptyset, \pi)\}$, and $\{(\emptyset, \emptyset), (\emptyset, \pi)\}$.

3.2 a) One is $(0, (1, 0))$. b) $(5)(5)(10)$.

3.3 a) $\{\emptyset, \{(c, e, a)\}\}$. b) $(\emptyset, \emptyset, \emptyset)$, $(\emptyset, \emptyset, \{a\})$, $(\emptyset, \{e\}, \emptyset)$, $(\emptyset, \{e\}, \{a\})$, $(\{c\}, \emptyset, \emptyset)$, $(\{c\}, \emptyset, \{a\})$, $(\{c\}, \{e\}, \emptyset)$, $(\{c\}, \{e\}, \{a\})$, and every bracket and parenthesis *is* important.

4.1 a) Let $x \in A \times B$. So $x = (a, b)$ for some $a \in A$, and some $b \in B$. Since $A \subseteq B, a \in B$. Since $B \subseteq C, b \in C$. Thus $(a, b) \in B \times C$.
Therefore $(A \times B) \subseteq (B \times C)$.
b) Consider if $A = \emptyset$.

5.1,2 All the flaws are coming from confusing 'an element of', notation $\in$, with 'a subset of', notation $\subseteq$. Phrases like 'contained in', 'is in', 'is part of' are vague and ambiguous, and can confuse both the reader and the writer.

5.3 There are of course, no flaws in part three. That is a valid argument. (Unfortunately it is based on the non-sense we concluded from the two flawed parts...)

6.3 All three.

Chapter 3 – Working with finite sets

1.1 a) $2^5 \leq |X \cup Y| \leq 2^6 + 2^7$. b) $0 \leq |X \cap Y| \leq 2^6$.

1.2 a) $2^{\binom{4}{2}4 \cdot 5}$. b) $\binom{2^4 4 \cdot 5}{2} + \binom{0}{3}$. c) 0.

2.1 The tenth: $10_{10} = 1010_2$ corresponds to b, d.

2.2 There are ten. $\{a, b\}, \{a, c\}, \ldots \{c, e\}, \{d, e\}$.

2.3 $999_{10} = 1111100111_2$ corresponding to $\{0, 1, 2, 5, 6, 7, 8, 9\}$.

3.1 $2 \cdot 5^4 + 5 - 5^3 - 1 - 1 + 1$.

3.2 $5^3 + 5^3 + 5^4 - 5^2 - 5^2 - 5 + 1$.

3.3 $26^{21} + 26^5 + 26^{13} - 26 - 26^9 - 26^3 + 26$.

4.1 a) $((1, 1, 1, 1, 1), 1, \{1\})$, $((1, 1, 1, 1, 1), 1, \{2\})$, $((1, 1, 1, 1, 1), 2, \{2\})$, $((1, 1, 2, 1, 1), 1, \{2\})$, $((2, 1, 1, 1, 1), 1, \{2\}) \ldots$
b) $|(A^5) \times A \times A| = |A^5| \cdot |A| \cdot 2^{|A|} = 3^6 2^3$.

4.2 a) $(((a, b), (a, c)), ((a, b), (a, c)))$, $(((a, a), (a, a)), ((a, a), (a, b)))$, $\ldots$
b) $((3^2)^2)^2 = 3^8$.

4.3 a) Puszat, Bozsey, ... b) 2^7.

5.1 None.

5.2 $2 \cdot 5^9 + 0 \cdot 5^8 + 2 \cdot 5^7 + 0 \cdot 5^6 + 1 \cdot 5^5 + 4 \cdot 5^4 + 1 \cdot 5^3 + 1 \cdot 5^2 + 4 \cdot 5^1 + 4 \cdot 5^0$.

5.3 Divide alternately by 3 and 2 and take remainders: $88 = 29(3) + 1$ so last character 2, $29 = 14(2) + 1$ so then b, $14 = 4(3) + 2$ so then 3 $4 = 2(2) + 0$ so then a $2 = 0(3) + 2$ so then 3.
Final result: $1a3a3b2$.

6.1 Read top to bottom, left to right:

abcd	bacd	cabd	dabc
abdc	badc	cadb	dacb
acbd	bcad	cbad	dbac
acdb	bcda	cbda	dbca
adbc	bdac	cdab	dcab
adcb	bdca	cdba	dcba

6.2 $8 \cdot 9! + 2 \cdot 8! + 1 \cdot 7! + 2 \cdot 6! + 2 \cdot 5! + 4 \cdot 4! + 2 \cdot 3! + 0 \cdot 2! + 1 \cdot 1! + 0 \cdot 0!$.

6.3 8,214,596,073; 8,214,596,307; 8,214,596,370; 8,214,596,703; 8,214, 596,730; 8,214,597,036; 8,214,597,063; 8,214,597,306; 8,214,597,360; 8,214,597,603; 8,214,597,630; 8,214,603,579; ...

7.1 $\delta\beta\alpha\gamma$.

7.2 $720 = 6!$, so $1(6!) + 0(5!) + 0(4!) + 0(3!) + 0(2!) + 0(1!)$ giving $bacdefg$.

7.3 0,123,968,457.

Chapter 4 – Formal logic

1.1 Natural language is slippery.

1.2 But are they all true?

1.3 The problem seems to be if Eddie only has a ruby...

2.1 Both are false.

2.2 Both are true.

2.3 Only all true or all false work.

3.1 Except for quibbling, all pass. The first two conclusions are true and the last antecedent is false.

3.2 a) The antecedent is false. b) The consequence is true.

3.3 b) Suppose $p \wedge q$. Then p is true and q is true. Since q is true, $q \vee r$ is true, as required.

4.1 We first show $[p \wedge (q \wedge r)] \Rightarrow [(p \wedge q) \wedge r)]$. Suppose $p \wedge (q \wedge r)$ is true. Then p is true, and $q \wedge r$ is true. Since $q \wedge r$ is true, both q and r are true. Thus $p \wedge q$ is true, and so $(p \wedge q) \wedge r$ is true. (Try now the second half.)

5.1 $((s \vee \neg r) \vee \neg q) \vee \neg p$.

5.2 $[p \wedge (r \vee \neg q)] \vee [\neg p \wedge \neg(r \vee \neg q)]$.

5.3 $[(\neg p \vee q) \wedge \neg(\neg q \vee p)] \vee [\neg(\neg p \vee q) \wedge (\neg q \vee p)]$ and many alternate forms.

6.1 $\neg(p \wedge q \wedge r)$.

6.2 $(p \wedge q \wedge r) \vee (\neg p \wedge \neg q \wedge \neg r)$.

6.3 $(p \vee \neg(q \wedge r \wedge s)) \wedge (q \vee \neg(p \wedge r \wedge s)) \wedge (r \vee \neg(p \wedge q \wedge s)) \wedge (s \vee \neg(p \wedge q \wedge r))$. There are many other ways to express these.

7.1 a) $(\neg p) \vee (\neg q) \vee (\neg r) \vee (\neg s)$ (short AND clauses). b) $(\neg p \vee \neg q \vee \neg r \vee \neg s)$ (just one OR clause).

7.2 a) $(p \wedge q \wedge r \wedge s) \vee (\neg p \wedge \neg q \wedge \neg r \wedge \neg s)$. b) $(p \vee \neg p) \wedge (p \vee \neg q) \wedge (p \vee \neg r) \wedge (p \vee \neg s) \wedge (s \vee \neg p) \wedge (s \vee \neg q) \wedge (s \vee \neg r) \wedge (s \vee \neg s) \wedge (r \vee \neg p) \wedge (r \vee \neg q) \wedge (r \vee \neg r) \wedge (r \vee \neg s) \wedge (s \vee \neg p) \wedge (s \vee \neg q) \wedge (s \vee \neg r) \wedge (s \vee \neg s)$ just using the distributive law.

7.3 a) You cannot have just one false. So $\binom{4}{2}$ terms $(\neg p \wedge \neg q) \vee (\neg p \wedge \neg r) \vee (\neg p \wedge \neg s) \vee (\neg q \wedge \neg r) \vee (\neg q \wedge \neg s) \vee (\neg r \wedge \neg s) \vee (p \wedge q \wedge r \wedge s)$.
b) Try $(\neg p \vee \neg q \vee \neg r \vee s) \wedge (\neg p \vee \neg q \vee r \vee \neg s) \wedge (\neg p \vee q \vee \neg r \vee \neg s) \wedge (p \vee \neg q \vee \neg r \vee \neg s)$. Why would that ever work?

Chapter 5 – Induction

1.1 a) True: Richard Feynman <u>was born after 1888</u>.
False: Albert Einstein <u>was born after 1888</u>.

1.2 a) True: 36 <u>is the sum of three distinct cubes</u>.
False: 35 <u>is the sum of three distinct cubes</u>.

1.3 It is small diagonal red "eight" in a sea of blue.

2.1 You get to choose n: Let $n = 1$, then $1^2 + 5 = 6 \cdot 1$, as required. (Where did the 1 come from? Work had to be done but that is not part of the proof.)

2.2 Let $n \in \mathbb{N}$. Consider two cases, n is even, ... (Now you are started.)

2.3 a) and b) are true by choosing $m = 0$. c) is false, since given n, you can take $m > n^3 + 1$.

3.1 The base case is $3 \cdot 0! > 0^2$.

The induction step requires that you prove, for any given $n \geq 0$, that $3 \cdot n! > n^2$ implies $3 \cdot (n + 1)! > (n + 1)^2$.

The induction hypothesis, for the given $n \geq 0$, is $3 \cdot n! > n^2$.

3.2 Hint: The fact that the statement is false has nothing to do what you *would do* to prove it.

3.3 P_5 must be true. P_{25}, P_{35}, and P_{45} must be false. P_{15} cannot be determined.

4.3 Since P_{15} is true, any predicate on an odd index greater than 15 is true by induction, so P_{55555} is true.

Since P_{14} is false, any predicate on an even index less than 14 is false, otherwise P_{14} would be true by induction, so P_0 and P_4 are both is false. About P_{2100} we can make no conclusion.

6.1 $\{0, 1, 3\} = \{0, 1, 2, 3, 4\} \cap \{0, 1, 3, 4, 5, 6, 7, 8, 9\}$.

6.2 Let the universe $\mathcal{U} = \mathbb{N}$. $T = \{2\}$. $\mathbb{P}$ is the set of primes. $E = \{n \in \mathbb{N}; n = 2k; k \in \mathbb{N}\}$. $\mathbb{P} \cap E \subseteq T$.

6.3 Define p_n to be 1 is $k - 1$ is evenly divisible by 5 and 0 otherwise.

Chapter 6 – Set structures

1.1 Hint: Notice that very few letters are related to many digits.

1.2 Hint: Notice that the specification does not stipulate that the two elements of A are distinct.

1.3 2^6.

2.1 $R = \emptyset$?

2.2 Why not the following? And why 13? Why do we need two lines at all?

$$f(A) = \begin{cases} |A| & \text{if } A \text{ is finite} \\ 13 & \text{if } A \text{ is infinite} \end{cases}$$

2.3 It is functional, one-to-one and onto. Can you show it?

3.1 Onto:

$$\binom{6}{6}6^{10} - \binom{6}{5}5^{10} + \binom{6}{4}4^{10} - \binom{6}{3}3^{10} + \binom{6}{2}2^{10} - \binom{6}{1}1^{10} + \binom{6}{0}0^{10}$$

3.2 Compute first $2^{(2^{(2^0)})} = 4$. Ordinary 4^4, all others $4!$.

3.3 $\binom{25}{16}9!$.

4.1 Hint: To show non-onto you may exhibit any violator.

4.2 Hint: Find two subsets with eight elements with the same maximum and minimum elements.

4.3 Hint: The target is an infinite set.

5.1 C_f consists of all integers except 0 and 1.

5.2 For instance $f(a) = f(d) = f(e) = f(g) = \emptyset$. and $f(b) = f(c) = f(f) = \{a, b, c, d, e, f, g\}$.

5.3 Hint: Besides the error of name calling, has the procedure assigned a value to *every* set? Why is that in issue?

6.1 There are plenty of reasonable arguments that it is uncountable. $\mathbb{R} \times \mathbb{R}$ contains $\mathbb{R} \times \{0\}$ starts one.

6.2 Use finite subsets of a countable set ...

6.3 A fun argument is to encode the numbers via placement and degree of the digit rises. A shorter one argues the numbers are all rational.

7.1 Reflexive, anti-symmetric, and transitive.

7.2 a) Anti-reflexive, anti-symmetric, and anti-transitive. b) All 6^2 arrows would be in the diagram.

7.3 Here's one:

8.2 Transitivity: Let $n \propto m$ and $m \propto r$. Then $n - m = 10k$ for some $k \in \mathbb{Z}$ and $m - r = 10j$ for some $j \in \mathbb{Z}$. Adding the two equations gives $n - m + m - r = 10k + 10j$, or $n - r = 10(k + j)$ so $n \propto r$, and the relation is transitive.

8.3 Here's one: $\succ$ isn't reflexive: 7 is not related to 7 since 14 is not divisible by 10.

Chapter 7 – Elementary number theory

1.1 You only need to sieve out primes 2, 3, 5, 7, and 11.

1.2 Don't use a calculator. Largest prime to check is 19.

1.3 $1776 = 2(888) = 2(8)(111)$ is one way to start.

2.1 Hint: $321 = 2 \cdot 123 + 75 \ldots 21 = 3 \cdot 6 + \boxed{3}$.

2.2 Start $988,887 = 1 \cdot 888,887 + 100,000$ then stop and note that $988,887$ and $888,887$ have the same common divisors as $888,887$ and $100,000$.

3.1 Give 7 Alexanders, and receive back 11 Bucephaluses.

3.2 $17(449) + (-72)(106) = 1$ can be found without calculator in less than 90 seconds.

3.3 There is no solution since $99 = 3^2 \cdot 11$, and $111 = 3 \cdot 37$.

4.1 Modulo 11: $6 + 7 = 13 \equiv 2 \bmod 11$.
$8 + 8 = 16 \equiv 5 \bmod 11$.
$2 \cdot 8 \equiv 5 \bmod 11$ is the same problem.
$1 + 2 + 3 + 4 + 5 + 6 + 7 + 8 + 9 = 1 + (2 + 9) + (3 + 8) + (4 + 7) + (5 + 6) \equiv 1 \bmod 11$.
$5280 + (65)(88) = 5280 + (65)(8)(11) = 5280 = 11(480) \equiv 0 \bmod 11$.

4.2 Additive inverse sum to zero: 0 and 0, 1 and 6, 2 and 5, 3 and 4.
Multiplicative inverses multiply to 1: 1 and itself; 2 and 4; 3 and 5, and 6 and itself.

4.3 Additive inverse sum to zero: 0 and 0, 1 and 7, 2 and 6, 3 and 5, 4 and itself.
Multiplicative inverses multiply to 1: 1 and 1; 3 and 3; 5 and 5, and 7 and 7 are the only pairs – weird.

5.1 From the Euclidean algorithm: $(1)(66) + (5)(-13) = 1$ The multiplicative inverse of 5 modulo 66 is -13. Then $5x = 3$ gives $(-13)(5x) \equiv (-13)(3)$ mod 13, so $x \equiv -39 \equiv 27$ mod 66.

5.2 15 mod 25.

5.3 Summary: $(1)(1) \equiv (2)(13) \equiv (3)(17) \equiv (4)(19) \equiv (6)(21) \equiv (7)(18) \equiv$ $(8)(22) \equiv (9)(14) \equiv (11)(16) \equiv (12)(23) \equiv (13)(2) \equiv (14)(9) \equiv$ $(16)(11) \equiv (17)(3) \equiv (18)(7) \equiv (19)(4) \equiv (21)(6) \equiv (22)(8) \equiv$ $(23)(12) \equiv (24)(24) \equiv$ mod 5.

6.1 $n = (11)(-2)(19) + (17)(3)(13)$.

6.2 All solutions are given by $5(-2)(19) + (2)(3)(13) + k(19)(13)$ for some $k \in \mathbb{Z}$.

6.3 $10(-2)(19) + (12)(3)(13)$.

Chapter 8 – Codes and cyphers

1.1 $10^{200} \equiv 1$ mod 11.

1.2 a) $2^{1776} \equiv 4$ mod 12. b) $2^{1776} = (2^8)^{222} \equiv 1^{222} = 1$ mod 17.

1.3 $2^{1776} = 2^{1770}2^5 2 \equiv 1 \cdot (-1)(2) \equiv -2 \equiv 31$ mod 33.

2.2

	a^0	a^1	a^2	a^3	a^4	a^5	a^6	a^7
0	0	0	0	0	0	0	0	0
1	1	1	1	1	1	1	1	1
2	1	2	4	1	2	4	1	2
3	1	3	2	6	4	5	1	3
4	1	4	2	1	4	2	1	4
5	1	5	4	6	2	3	1	5
6	1	6	1	6	1	6	1	6

The 6th and 7th columns are just what are predicted by Fermat's Little Theorem.

2.3 b) $2^{1710} = (2^{1700})(2^{10}) = (2^{100})^{17}2^{10} \equiv 1 \cdot 1024 \equiv (10)(101) + 14 \equiv 14$ mod 101.

3.1 The multiplicative decoding key is the multiplicative inverse of 7 modulo 26 which is 15.

3.2 Hint: The encoding and decoding keys are established modulo 10 but the encoding and decoding are done modulo 11.

3.3 For exponential, the numbers not coprime to 10 are $\{0, 2, 4, 5, 6, 8, 10\}$.

4.1 $\delta = 77$.

4.2 $(108, 141, 166)$.

4.3 Hint: You need the multiplicative inverse of 67 modulo $(3 - 1)(101 - 1) = 200$; which is 3.

5.1 $5280_2 = 1,0100,1010,0000$, a 13 digit binary number, so the Euclidean algorithm should take at most 26 steps by our bound.

5.2 $\lim_{n \to \infty} \frac{500n^5 + 1000n^3}{n^6} = \lim_{n \to \infty} \frac{500}{n} + 1000n^3 = 0 + 0 = 0$.

5.3 $\lim_{n \to \infty} \frac{2^n + 3^n}{5^n} = \lim_{n \to \infty} (2/5)^n + (3/5)^n = 0$.

6.1 $1776^{(2^9)} \cdot 1776^{(2^7)} \cdot 1776^{(2^4)} \cdot 1776^{(2^3)} \cdot 1776^{(2^1)}$.

6.2 $7^{22} \equiv 49 \bmod 100$.

Chapter 9 – Graphs and trees

1.1 The number of edges in a simple graph on n vertices is bounded by $\binom{n}{2}$, since we have to choose two endpoints for every edge. Complete graphs achieve this bound. K_5 has 10 edges. K_n has $\binom{n}{2}$ edges.
Since a graph in general can have multiple edges there is no upper bound.

2.2 The total cost of the resulting road network is $240 + 270 + 300 = 810$.

2.3 Hint: By Cayley's formula, there are exactly $4^2 = 16$ such trees.

3.2 Base case: $k = 0$. We have $T(2^0) = T(1) = 0 \leq 0 \cdot 2^0$, and the inequality holds.
Induction step: Let $k \geq 0$, and assume that $T(2^k) \leq k \cdot 2^k$. We want to show that $T(2^{k+1}) \leq (k + 1) \cdot 2^{k+1}$. We have $T(2^{k+1}) \leq 2T(2^k) + 2^{k+1}$. Using the induction hypothesis, we have $T(2^k) \leq k \cdot 2^k$ and therefore $T(2^{k+1}) \leq 2T(2^k) + 2^{k+1} \leq 2k \cdot 2^k + 2^{k+1} = k \cdot 2^{k+1} + 2^{k+1} = (k + 1) \cdot 2^{k+1}$, which completes the induction step.

4.2 If G is bipartite, it only contains cycles of even length, hence is triangle-free, and we can use Theorem 9.15. We could also provide spanning tree argument and "quadrangulate" the $(2n - 2)$-gon by $n - 3$ "diagonals".

5.1 The graph has four vertices of odd degree. Therefore, by Theorem 9.19, it has no Eulerian trail.

6.1 Consider a traceable tree T and let P be a Hamiltonian path in T. Then every edge of T must belong to P since otherwise T would contain a cycle. Therefore, the only traceable trees are paths (that is, trees in which all vertices have degree at most two).
As every tree is acyclic, there are no Hamiltonian trees.

Index

Printed in the United States
by Baker & Taylor Publisher Services